# TIGER JUNGLE

For Ruth, Abbie, Sammy and for my parents
who first inspired and encouraged my passion for wildlife

www.tiger-books.co.uk | www.chevronpublishing.co.uk

# TIGER JUNGLE

## The Epic Tale of Bandhavgarh

Written and Photographed by Iain Green

## Acknowledgments

It has been ten years since I first went to Bandhavgarh and during that time I have been greatly assisted by so many people.

In India, my thanks go, first and foremost, to my friends Kay and Satyendra Tiwari, guides and naturalists par excellence. They have taught me so much about the jungle and its daily drama. Only through their quiet fieldcraft in the forest and then an analysis of the day's events once back at camp, have I been able to build this portrait of Bandhavgarh. I would also like to thank Satyendra's brother, Hariam, and his family, Sonu, Munmun and Mahi, who always make my stay at the camp so pleasurable – papayas, mangos and angelic smiles.

I have made many other friends in Tala village who have helped shape my appreciation of the forest: a big thank you to Rama Shanker Yadav for his work in the early years. To Rajesh and all the forest guides. Kuttapan, Mantu, Dheram and the other mahawats. The park directors, rangers and beat guards who make Bandhavgarh what it is. Dhruv Singh, Lakshmi and Lucy at Churhat Kothi. And to all my other friends in Tala. Also in India I would like thank, Alphonse Roy, Ragu Chundawat and the various NGOs working to protect tiger habitat. And finally to Mr Singh and his family in Bharatpur.

In the UK, I must extend thanks to David Shepherd, OBE, FRSA, FRGS. I would also like to express my sincere thanks to everyone at the Environmental Investigation Agency, in particular Debbie Banks - a tireless campaigner for tigers; Andy Fisher at Metroplitan Police Wildlife Crime Unit and all the other organisations working to combat poaching and habitat loss in India. And special thanks to Dr Debbie Pain and the RSPB for providing me with the latest information on India's vulture crisis. I sincerely hope that they are successful in averting another wild catastrophe.

While researching the history of Bandhavgarh, I have been greatly assisted by Dr Terence Walton who gave me a unique insight into the area before it was a National Park. His memories and old images have greatly enhanced my portrait of the jungle. I have also been helped by Madeline Baglow, who kindly allowed me to reproduce the previously unseen images of Mohan.

In turning this book into a reality I must thank my publisher, Robert Forsyth, for 'agreeing to do it all again' and his constant support. Actpix in Wales for expertly scanning all my transparencies. Mark Nelson for creating a beautiful book. And Sally Forsyth, who started it all.

Also to Richard Squibb at Vine House Distribution. Tim Harris and everyone at my superb agents PhotoShot/NHPA. Also to Laura and Nick Barwick, Clare Fisher, Theo Allofs, Kim Sullivan, Paula Rodney, and all my family for their support.

And lastly to my wife, Ruth, for her never-ending encouragement and constant support. And for putting up with my unsocial hours of work.

First published in Great Britain in 2007 by Tiger Books (UK)
an imprint of Chevron Publishing Limited
Friars Gate Farm, Mardens Hill, Crowborough
East Sussex, TN6 1XH, England

tiger.books@chevronpublishing.co.uk
www.tiger-books.co.uk
Project Editor: Robert Forsyth
Cover and book design by Mark Nelson


ISBN (10) 0 9543115 2 3
ISBN (13) 978 0 9543115 2 0
Printed in Singapore

# CONTENTS

# Introduction

ON my studio wall hangs this beautiful study of an Indian leopard, drawn in 1881 by my great-grandfather, a professional artist. My family connection to India is a strong one that spans many centuries – the passion for the country continues with me. The inspiration for my love of wildlife was my father and he himself spent his early childhood in South India. He lived in Bangalore, where my grandfather was Commissioner of Police, until India gained its independence in 1947.

Forty years after my family returned to England, I made my first visit to India, backpacking round this vast and diverse land. Such a contrast from anything I had encountered before, India was an intense sensory experience – hitting both the high and the lows. Like a young child at their birthday party, my time in India travelled a fine line between ultimate excitement and total trauma. Exhausted from too much travelling, the trip didn't end well; I contracted malaria and being a little headstrong, I vowed I would never return.

However, a decade later and a lot more mellow, I went back to India to visit the famous bird wetlands in Bharatpur - and so began my professional wild photography career. Photographing in the bird paradise at Bharatpur re-ignited my fire for the Subcontinent and the following year I went again, but this time to a place called Bandhavgarh, where I hoped that I just might see a tiger.

On my first day tracking in the forest, I was rewarded by the discovery of a magical jungle, but saw only pugmarks of tigers in the soft sandy soil. Then on the second evening, a tigress, known as Bachchi, fleetingly crossed the forest track in front of my jeep. The next day I found her three cubs playing in an open area of Sal forest. It was an incredible encounter that resulted in some of my favourite photographs. Since then I have returned regularly to the tiger jungles to track the tigers and it is these three brothers that have become the focus for my unique photographic study. I believe in authenticity in wildlife photography – all the images inside this book are of wild animals and none have been manipulated in any way.

Like any experienced photographer I draw on the knowledge of local experts and on my first visit I struck lucky. Bandhavgarh's unofficial naturalists, Kay and Satyendra Tiwari, have become great friends, hosts and guides to me and without them this book and my studies would be so much the poorer. They have an amazing understanding of the tigers' behaviour and of the other jungle inhabitants – their knowledge of the forest butterflies is second to none. Kay is an accomplished wildlife artist (the map is her creation) and Satyendra is an award-winning guide and wildlife photographer – some of the photos on pages in Chapter 7 are his. Living on the edge of the forest they track the tigers most days and over three decades have gained an unparalled picture of the jungle. Bandhavgarh also has a hugely absorbing human history and I have been lucky enough to talk to people who knew the area before it was a National Park and who were directly involved in itscreation. Their knowledge adds to my understanding of this epic place.

Sadly, tigers are in trouble across their entire range. Despite decades of international conservation campaigns, tiger numbers are near an all-time low. Almost every week there is a new story of poaching, habitat loss or the far-reaching effects of climate change. It is easy to be overwhelmed and assume tigers will become extinct, but this is not a foregone conclusion. Decisive action taken now, directed in the right place, can save the tiger. Tigers don't need captive breeding programmes, they simply need properly protected habitat – Bandhavgarh clearly shows how well they breed in the wild. There needs to be solid commitment from governments by supporting the scientists and habitat managers working in the field, but we can all make positive steps to protecting the tigers and their habitat. Small actions taken by lots of people are often the best. Support tiger conservation organisations such as those listed at the back of this book. Keep the issue at the top of the political agenda by writing to politicians, both at home and in India or Asia.

In our own lives we can try to reduce our environmental impact. Climate change affects every person and place – wildlife photographers are no exception. Flying abroad is one of the biggest environmental threats and one that I now cannot ignore. Regularly jetting off around the world to photograph endangered animals is a clear inconsistency. Though I will continue to follow the lives of the tigers, I have made the logical, but difficult decision to greatly reduce my long-haul travel. This is why some of the latest images in the book are Satyendra's. We all have a duty to do what we can to help ensure that our huge footprint does not destroy the planet's fragile wild habitats.

The tiger jungles of Bandhavgarh are one of the world's treasures – enjoy.

**Iain Green, August 2007**

# Foreword by David Shepherd, OBE, FRSA, FRGS

I was only too delighted to accept the invitation from Iain Green to write a foreword for this magnificent book with its wonderful photographs of the poor, beleaguered tiger. I say this because I welcome any opportunity to speak up for this glorious animal which, in my view, is the most beautiful in the world.

We all know that the survival of the tiger is critical. The numbers vary; the 'official' estimate from the Indian government is just over 3,000, but I consider it is far more serious than that, the number possibly being only 1,500. I have had the thrill and pleasure of seeing tigers in the wild on a number of occasions and I am always left in awe at their sheer beauty. I am also in despair when we are trying to get the message through to those in power who could do so much more. I am always keen to speak bluntly about issues that affect all living creatures on this planet; when we were visiting the Taj Mahal, the Indian official with us was expressing deep concern about the future of the building because of pollution. "You can always build another one," I said. He was aghast at my suggestion, but I was simply making the point that the tiger is God-given and unlike a building, could be lost forever.

The more attention that is drawn to the fate of the tiger now, the better, and I am sure this book will play a substantial part in this, illustrating that these animals have a right to survive in their natural environment.

David Shepherd

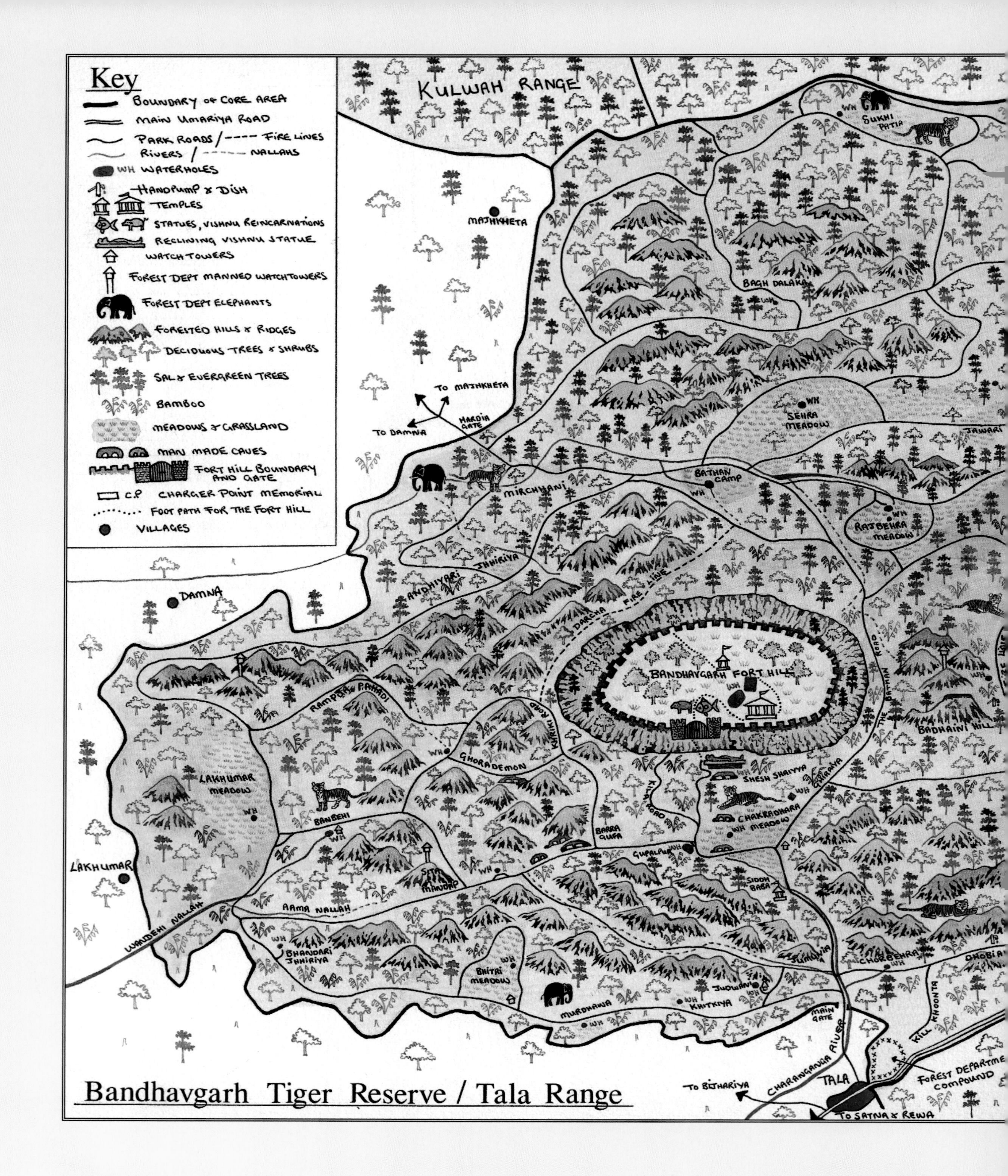


Bandhavgarh Tiger Reserve / Tala Range

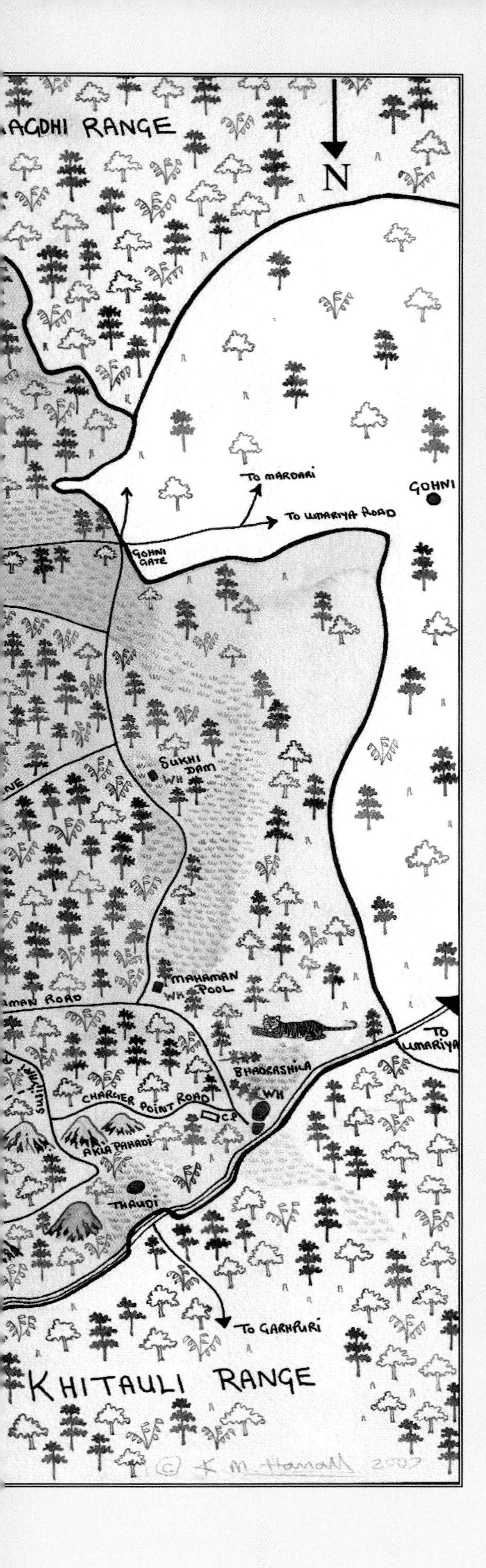


Situated in the heart of India, Bandhavgarh lies on the northern edge of the Satpura range, in the state of Madhya Pradesh. Just north of the Tropic of Cancer, at 23° 50" North and 80° 50" East, Umaria is the nearest town and rail station to these tiger jungles. The forests have three clearly defined seasons. Winter lasts from November to February, at the coldest times the temperature gauge may almost hit zero. The climate begins to warm in March and the hot season peaks at more than 45°c in May and June. At the end of June, the monsoon rains arrive, instantly killing the heat but bringing high humidity. The average rainfall is 50 inches, though in recent years it has been lower. By September, or occasionally early October, the rains end.

This is a pictorial map of the Tala Range in Bandhavgarh Tiger Reserve, painted by wildlife artist and naturalist, Kay Hassall Tiwari. Kay and her husband Satyendra, are my guides and hosts during my stays in Bandhavgarh.

The tiger illustrations on the map indicate rough locations of female territories in 2007 (see family tree on page 174 for details of name codes).

Pyari (S1) - Chrakradhara tigress

Reshma (S2) - Banbehi tigress

Chameli (K3) - Sukhi Patia tigress

Durga (P6) - Badhaini tigress

Lakshmi (P4)- Chorbehra tigress

Indrani (P5) - Badrashila tigress

Tulsi (R10) - Milchaini tigress

# The Place of Legends

## 100 AD-1998

For more than a decade Sundar has lived in the lush sal forests of Bandhavgarh in Central India. Also known as B2, he has been the territorial male of the National Park's core area, the Tala range, for the past seven years. As Sundar walks through the jungle, the habitat made famous by Kipling's 'Jungle Book', his daily journey takes him over the remains of lost civilisations, and below the walls of a fort belonging to epic Indian folklore. His pugmarks imprint on the soft sandy soil, marked by tigers for so many years. The intertwined history of Bandhavgarh and tigers is a long one – from Maharajas and Mohan to temples and tourists.

But the tigers' world is in turmoil; latest estimates suggest that the entire worldwide wild population may be as low as 3500 animals and decreasing. Sundar has enjoyed a successful life, fathering many litters of cubs, but for how long will his descendents continue to leave their mark in this epic land.

Pronounced 'Bandho-Garh', literally 'the brother's fort', Bandhavgarh National Park takes its name from the central plateau that towers almost 400 metres above the surrounding meadows. Though tigers may be the draw to this jungle it is the dominance of the mighty hill that makes the first impression on visitors as they enter the forest. Mythology and folklore are a strong part of present-day India. Dating back to the time of the Hindu epic, the Ramayana, ancient legend tells how Lord Rama gave the fort to his beloved younger brother Lakshman. Rama, the seventh incarnation of the Hindu god Vishnu, the protector, had just returned from his battle in Lanka with the demon king, Ravana. Hanuman, the monkey god, and his followers helped Rama win the war by building a bridge between the mainland and the Island of Lanka. According to the legend, it was Hanuman who built the fort. Upon receiving the gift from his brother, Lakshman became known as Bandhavdhish – Lord of the Fort.

Factual history and folklore easily merge in Bandhavgarh, especially on the fort itself. Exploring the fort, there is the excitement of being 'Indiana Jones' discovering the hidden rock-hewn sculptures and ruins from the past millennia overgrown by vegetation. Walking from the stone water tank at Sheshshaiyya, a steep track snakes upwards among the jungle clad cliffs, through the fort gates, past ancient religious carvings to the hill top temple and ultimately a stunning view from the platform known as the Maharaja's Seat. The platform was, in fact, the deck from where a ceremonial canon was fired – the last time being in 1943 on the occasion of Maharaja Martand Singh's wedding. The fort is one of the few areas of the National Park that can be explored on foot – tigers are less frequently seen up there. However, they do still visit the area, adding a little trepidation to tourists or pilgrims that make the climb. Sundar rarely visits the plateau top but does bathe in the frog-filled water tank at Sheshshaiyya. More than one human visitor has had to carefully back off, on finding him only metres away relaxing in this enchanting pool.

Carved from the natural sandstone rocks in the 10th century during the Kalchuri dynasty, the Sheshshaiyya statue shows Vishnu, the Hindu god, reclining on Sheshnag, the king of snakes. The 11 metre-long carving lies at the back of the spring fed pool and it is from Vishnu's feet that the river Charanganga (literally Ganges from the feet) has its source and origin. Flowing through the forest and meadows, the Charanganga is a very important source of water for Bandhavgarh's wildlife.

Climbing further up the hill through the imposing fort gates at Kam Pol, another magnificent sandstone statue appears through the trees. Standing seven metres tall, Varaha (the boar) is the third incarnation of Vishnu and is shown in its part human form. Once on top of the plateau there are carvings of the other avatars of Vishnu, including a tortoise, fish and Lord Krishna. Having all ten incarnations of Vishnu, the fort is hugely significant to Hindus and many followers of Vishnu make the pilgrimage on the Birthday of Lord Krishna in August, and in April, on the birthday of Lord Rama,in order to worship and leave offerings.

**Sheshshaiyya**

*Jungle is a word often used to describe the tropical rainforests of Africa and South America, but it is actually derived from the Indian word 'jungal' meaning desert, wasteland or forest.*

These jungles of Central India inspired Rudyard Kipling to write the Jungle Book in 1894. Though strictly set in the Seoni hills at nearby Pench National Park, the forests and animal characters of Bandhavgarh are similar. As well as tigers, Bandhavgarh is home to Baloo the Sloth bear, wolves, Kaa the royal python and the Bandar-log. Many of the characters names are actually based on the Indian animal's name - Bhalu is the local word for bear and Sher means tiger. The forest's lively groups of hanuman langurs are clearly the inspiration for Kipling's Bandar-log, or monkey people.

Secret tunnels, superb natural fortifications and high elevation have given the upper hand to successive dynasties for whom Bandhavgarh has been a place of rule for more than a thousand years. Covering an area of 582 acres, the plateau is large enough to support a whole community – there are the remains of dwellings, school rooms, numerous temples and twelve massive man-made tanks cut into the rock to provide drinking water for the residents.

Old stories suggest that on the southern side of the fort there is a secret tunnel, to allow supplies and people to come and go undetected. Using pre-arranged calls a rope would be lowered through the tunnel to the forest below. Personnel and goods would then be hauled up or lowered and the rope hastily raised, leaving no visible trace.

Between the third and ninth centuries A.D., Bandhavgarh was the home to a number of dynasties including the Vakatas, Sengars and Balendu Khastriyas

The Kalchuri Clan took rule of the area in the 10th century A.D. and during the next three centuries they were responsible for carving the many rock sculptures of Vishnu, including the one at Sheshshaiyya.

In the late 13th century, Karandeo, a Baghel Rajput from Gujarat, was given Bandhavgarh as part of a dowry when he married a Kalchuri princess. Not long after, fighting forced the Baghels to leave their Gujarat homeland and many made the journey to Bandhavgarh where they established their new capital. Coincidentally the Baghel clan, or Vaghelas, take their name from the village in Gujarat, Vaghela, which means Tiger's Lair.

Over the next three centuries the Baghels enlarged their territory, until 1597 when the accession of a minor, Vikramaditya, to the throne caused a period of disturbance. Akbar, the famous North Indian Emperor, who had a soft-spot for Bandhavgarh, intervened and after an eight-month siege forcefully quelled the troubles. It is believed that, while she was pregnant with Akbar, his mother sheltered in Bandhavgarh fort, a safe place to hide while India was experiencing an Afghan invasion.

By 1617 the boundaries of the Baghel territory had extended so far that it made sense to move their capital to a more central location, the town of Rewa, 120km north of Bandhavgarh. With the King, his armies and people gone, the deserted Bandhavgarh gradually returned to jungle – animals and plants slowly took over the former city.

**The Tala Range of Bandhvagarh in 2002 viewed from the Maharaja's Seat.**

With the potential for prime populations of tiger and prey, it was clear that these new forests would make superb hunting territory and were soon declared a Shikargah, or hunting preserve for royalty. Now known as the Maharajas of Rewa, the Baghel kings had unknowingly given birth to one of the wildlife jewels of India – though the wildlife was still far from safe. More than three hundred years of private hunting by the Maharajas and their guests must have taken their toll on the animal populations. Tigers were the top prize and it was considered auspicious for the royal hunters to each shoot more than 100 tigers! One Maharaja shot 83 in a single year.

The last ruler of Rewa, His Highness Shri Martand Singh, succeeded his father as Maharaja in 1946, just in time to see India become independent the following year. The princely state of Rewa was then merged into the Vindhya Pradesh Union (now in Madhya Pradesh). Though India's Maharajas lost most of their power after Independence, they retained their titles and some privileges for a further 25 years; a period of huge significance for Bandhavgarh. Legal hunting was banned and Maharaja Martand Singh, a widely respected man, was responsible for two contrasting tiger legacies.

**The view from the Maharaja's Seat in 1967.**

Photographs reproduced with the kind permission of Madeline Baglow.

*Contrary to popular belief white tigers are not albino or a separate endangered species and they have no place in modern-day conservation. All the white tigers in captivity carry a recessive gene and are a mutation of the normal coloured Bengal or Amur tiger.*

**Mohan - Father of the white tiger**

In May 1951, the Maharaja of Rewa, received news of a white tiger cub in the jungles near Bandhavgarh. In the early 1900s there were occasional records of white tigers; indeed the Maharaja shot one himself in 1947. Wanting to capture one, Maharaja Martand Singh arranged a hunting party and soon found the mother and cubs. The mother and normal coloured siblings were shot by the Maharaja of Jodhpur, but Mohan (meaning 'the enchanter') was spared. The following day he was trapped in a cage and transported to Govindgarh near the Maharaja's palace in Rewa,

*The famous captive white tigers have their origins in Bandhavgarh, but though they have become popular tourist attractions in zoos and parks, it is because of their striking looks and not for their conservation importance.*

where he was kept in a courtyard (as these rare photos taken in 1964 show). Early on, the Maharajah made attempts to sell his unique cat, advertising it in the London and New York *Times*. However, at $10,000 no zoo or private collector wanted to buy, so the Maharajah decided to try and breed more white tigers.

The next year, a normal coloured tigress, known as Begum was captured and the two tigers mated. None of the cubs from three successive litters were white and all bar one were sent to other zoos. The Maharajah was not aware that both parents must carry the recessive 'white' gene to produce a white cub. Frustrated by the lack of success, the Maharajah decided to try and mate Mohan with the white tiger's own daughter, Radha. A carrier of the recessive gene, in 1958 she gave birth to four white cubs. Subsequent pairings of Radha and Mohan produced 13 white cubs. One of the daughters from the first litter, known as Mohini, was given to President Eisenhower at the White house for the US National Zoo in Washington. Mohini was mated with her normal coloured uncle and half-brother and also her own son. The resulting white cubs are the ancestors or North Americas' captive white tigers. Some captive white tigers also have an Amur (often called Siberian) ancestry. The famous American white tiger known as Tony, has an Amur Grandfather who was paired with a Bengal tigress. She produced two cubs, a brother and sister, who were then bred together resulting in Tony.

The lineage of white tigers is complicated, but does serve as a clear example of how inbred and cross-bred (Amur with Bengal) all captive white tigers are. The inbreeding has also led to a number of birth defects including cross-eyes, a weak immune system and spinal problems.

The father of the world's white tigers, Mohan died in 1969, at almost twenty years old.

Photographs reproduced with the kind permission of Dr. Terence Walton

**Maharaja Martand Singh (right) and Dr Terence Walton (centre) meeting wildlife officials during the late 1960s.**

## A vision of Bandhavgarh

After independence the Sub-continent entered a new era. Times had changed and so had attitudes to wildlife in India. Thanks to pioneering Indian naturalists, such as Salim Ali and E P Gee, the importance of conservation was starting to dawn. There was a growing realization that decades of indiscriminate hunting had taken a heavy toll on the country's tigers and these big cats were under serious threat. Action was needed immediately to save the tiger from the same fate as the Indian cheetah – the last known cheetah was believed to have been shot in 1947, not far from Bandhavgarh.

Like many others, His Highness Martand Singh Maharaja of Rewa wanted to do something for wildlife; enthused by a chance meeting he did the best thing possible.

On his way back to England from New Guinea, where he had been studying Birds of Paradise, Dr Terence Walton travelled to Rewa hoping that he might see Mohan, the now famous, white tiger. Within days of arriving and meeting each other, His Highness and Dr Walton were like old friends. Dr Walton subsequently became Equerry to the Maharajah – working for him for many years. Together they developed plans to create a wildlife sanctuary in the forests at Tala. With the Maharajahs will, these ideas became reality in a very short time. Grazing and hunting were stopped. However, bamboo collecting from the forest was the greatest problem and much more difficult to tackle because of the plants used by the powerful local paper industry.

May the day be not too distant when we shall see the wild life in dear old Bandhogarh roaming free, happy and unmolested in its rightful domain —!

With kindest regards.

Yours very sincerely —

Martand S.

Lengthy negotiations with Indira Gandhi's government ensued. At the same time, parliament was initiating land reforms that would hit landowners. In a mutually beneficial arrangement, the Government used its might to stop the bamboo collection and in 1968 the Maharajah of Rewa donated 105sq km of land at Tala to the state to become a National Park, although the fort plateau was not included in the donated land. Dr Walton maintains that it was one of the fundamental conditions that the fort remained the property of the Maharajah and Royal Family, accessible only with his permission. However, in recent years this has become a major dispute and at present the state authorities control access.

**A personal letter from the last ruling Maharajah of Rewa (above), His Highness Shri Martand Singh to his Equerry Dr Terence Walton.**

*At night the Maharajah and Dr Walton would make forays into the jungle on the back of an elephant called Gungawatti. Using a lamp wired to an old car battery they would explore the forest, getting surprisingly close to some animals like the Gaur, or Indian Bison. Sadly Gaur have not been seen in Bandhavgarh in the last decade.*

**The Kam Pol gate at the fort in 1967.**

In 1971 Indira Gandhi finally abolished the privileges and titles of India's Maharajas, but by this time the former Maharaja of Rewa, Martand Singh, had set in motion the creation of one of India's jewels.

In 1982 a further three ranges, Kalwah, Magdhi, and Khitauli were added, bringing the park's total area to 448 sq km. After more than three centuries of royal hunting, Bandhavgarh became a fully protected Project Tiger Reserve in 1993 when the neighbouring Panpatha Wildlife Sanctuary (245 sq km) was included. With the surrounding Reserve Forest or 'Buffer Zone', Bandhavgarh Tiger Reserve encompasses an area of 1161 sq km

*In the national park's early days, attempts were made at reintroducing Blackbuck to Bandhavgarh, bringing the antelope all the way from Calcutta to the fort plateau. The reintroduction was a total failure – though the fort leopard, who ate most of the Blackbuck, benefited.*

MARUTI

Tala is where it all began for the National Park and today this range continues to be at the heart of Bandhavgarh – geographically, historically and for tiger study. Though all the other ranges still contain villages, access to them for tiger study is severely restricted. Conversely, Tala is the only range fully accessible to tourists or naturalists and is now without any villages. When the Tala range became a National Park the villages were moved outside the boundary – the community at Bathan was the last to be relocated in 1976. Today, great swathes of grassland exist where there were once agricultural fields and dwellings. Aside from forest rangers, tourists are the only humans watching over the meadows. More than ten thousand Indian and foreign tourists, visit the park each year, eager to catch a glimpse of the elusive tiger.

# Tiger Jungle

Cascading from the fort plateau, the forests and streams plunge down the steep cliffs towards the tiger jungles of Bandhavgarh's Tala range. At the heart of Tala is Chakradhara meadow and in 1997 these grasslands and surrounding forests were the territory of probably the best-known wild tiger in the world, Sundar's grandmother, Sita. Together with her mate, Charger, she became famous overnight when a two-year photographic study for *National Geographic* hit the news stands. Front cover images of Sita caring for her young cubs, allowed the world a glimpse of wild tigers rarely seen. These images inspired me to visit Bandhavgarh and now ten years later I am still studying the tigers. The *National Geographic* cover shot shows Sita carrying one of her tiny cubs from one hiding place to another. Over the last decade this six-week old cub has become an eleven year-old tigress, known as Reshma or the Banbehi tigress, who has raised four litters herself.

Born in September 1996, Reshma and her sister Pyari are 21 months old and beginning to move independently of their mother. Already hunting successfully, they will become totally self-reliant over the next few months; cubs normally disperse between eighteen months and two years old.

With a group of wild boar in her sights, Reshma approaches, walking along the edge of the patchy meadow. The short dry grass offers no camouflage to her striking golden body, but once among the sal trees her stripes mingle well in the contrast of shade and streams of bright light. Further into the forest and nearer to her quarry she becomes invisible. The early morning jungle is quiet, just the occasional bird-call breaks the silence; the clucking of the red junglefowl, ancestor to the domestic chicken, is the most distinctive. Then the peace is interrupted by a mêlée of alarm calls, dashing spotted deer and scattering wild boar. A few minutes later Reshma appears briefly, crossing a small forest clearing with a wild boar piglet in her mouth! She will eat this time, but tigers are normally only successful once every ten or twenty stalking attempts.

Sita's male cub does not have the independent urge of his sisters and his mother has had enough. Affectionate gestures from her son definitely do not work – Sita's maternal warmth has been replaced by aggression, as she growls at her cub. Sita is sixteen years old and has given birth to six litters, an impressive grand total of 18 cubs. Some 'experts' question whether she really had so many cubs; however, Satyendra, with whom I work in Bandhavgarh, and many of the forest staff have seen all six of the litters. One of the oldest recorded tigers in the wild, Sita has been an incredibly successful mother, but she is now very tired and needs her son to leave.

Born in 1982 in an area known as Sita Mandap, Sita has been the 'Queen of Bandhavgarh' for more than a decade. Her first two litters were fathered by a male known as Barka, the then territorial male. Barka was deposed in 1991, by Charger, who became Sita's mate and together they ruled the Tala range for the next seven years, producing four more litters. Adult males do not

normally have any involvement with the cubs but Charger developed a special bond with his last male cub. Charger would greet his almost adult son and even seemed to reluctantly participate in the young tiger's playful behaviour.

These images were taken in June at the tail end of the hot season. For the past few months, temperatures of more than 40 deg C have baked, dried and a put strain on the National Park – the forest and all its inhabitants are waiting for rain and a break in the heat. The monsoon season should last for three months, but amounts of rain vary from year to year. In the wettest monsoons the park changes beyond recognition, lush verdant growth sprouts from every possible site, turning the forest a vibrant green. Tracks are washed away rendering the forest all but inaccessible; the daily lives and behaviour of the jungle's animals must change to survive.

Sita's male cub left the range during the rains to seek out a territory in another part of Bandhavgarh. It was also Sita's last year in Tala. Though rumours were rife that poachers had killed a famous tiger, a story that even made international headlines, it is perhaps likely, that Sita died of natural causes. Sixteen is quite old for a wild tigress and the monsoon rains would have inevitably brought extra survival challenges in Sita's daily life. Her remains have never been found, but Bandhavgarh is a large land with many inaccessible ravines and caves – the sort of places a tired tiger might go to rest. As is true with so much in wildlife watching, we will probably never know the complete story. However, it is now clear that she left a superb legacy – three daughters, Reshma, Pyari and from an earlier litter, a tigress known as Bachchi. During Sita's last summer in Bandhavgarh, Bachchi, like her mother, was also raising her own trio of cubs.

# Passage to Independence

## 1998-1999

Bachchi's boys play and explore the parched jungle in the relative cool of a summer dawn – it is probably only 30 deg C! While their mother is away hunting, Raj and Barra Larka practice marking their territory on a sloping sal tree. Sundar watches his brothers, from nearby.

As the day warms, the oppressive heat quickly kills all energetic behaviour. Regular midday temperatures of more than 45 deg C make the dry season an exhausting time for the three young tigers. Living on the other side of the Tala range to Sita, under Badhaini hill, it is already clear that Bachchi has learnt well from her mother. She has given her sons the best start they could have had. Bachchi's first litter, are now just over a year old and growing up fast – the brothers are showing signs of independence at a much younger age than Sita's son.

The long hot summer has scorched the vegetation; crisp brown sal leaves, carpet the sandy forest floor. The annual rains will soon bring some relief and it will not be long after the reviving monsoon passes, that the brothers take their first steps to self-reliance.

Bachchi, meaning 'little girl' in Hindi, grew up alone – the sole surviving cub from Sita's fourth litter born in 1994. Her only sister was born disabled and tragically could not keep up with her mother as she moved through the jungle. At ten months old, the cub was found wandering alone by forest rangers. Clearly something was wrong, but it took a little observation to discover that the young tigress was blind and partially paralysed. Easily distracted by jungle sounds and sometimes following human noises, the cub was increasingly in danger. The rangers considered raising her by hand but that would ultimately condemn her to a miserable zoo existence, so they had to let nature take its course. Inevitably, one night when out on patrol, they found the cub's body lying on the edge of a meadow – the rangers took her body away. Sadly, this was before Sita had discovered her dead daughter and she returned to the meadow at the same time every evening for four or five days desperately calling her cub. Confused, she could smell her daughter but could not find her; Sita would never know what had happened.

Local people, who witnessed this traumatic event, said it changed the way Sita behaved. She became increasingly secretive and kept her remaining daughter, Bachchi, well hidden. Over the next year, Sita concentrated her energies into teaching Bachchi all the skills needed for becoming a wild tigress and a successful mother. Learning from her own upbringing, Bachchi kept her first litter of cubs well hidden; the three male cubs were initially seen when they were already eight months old.

Born in April 1997, Raj, Sundar and Barra Larka have led a secluded life so far – helped significantly by the inaccessible habitat around Badhaini hill, an area at the core of Bachchi's territory. However they are now very inquisitive teenagers and while their mother is away they can explore as they please. For Raj, any new feature of his environment is worth closer study - be it a snake, a twig, or jeep.

Even for carefree young tigers like Raj and Barra Larka, the intense summer heat is very draining. Energetic behaviour is definitely not on the menu.

The cubs share their tinder-dry jungle playground with many other young animals. And, if given the chance, these fawns would certainly be on the brothers' menu. Sambar deer and chital – also known as spotted deer – are the tigers preferred item of prey in Bandhavgarh; although adult tigers would normally hunt fully grown deer.

Bandhavgarh is home to two species of primate, the largely ground-dwelling rhesus macaque and the more arboreal Hanuman langur or leaf-monkey. Though elusive in Bandhavgarh, rhesus macaques are common across India – even living in the heart of cities, where they feed significantly on discarded human food. The salt, sugar and fat-rich diet has altered the behaviour of these urbanite macaques and they have become larger, much more bold and at times aggressive.

Unlike their city cousins, Bandhavgarh's rural macaques are small shy monkeys. Often encountered on the edge of the jungle where open grasslands meet the sal forest, they are a delight to watch but become easily agitated. Quiet observation reveals the subtleties of their social structure, the mutual grooming between females, or the whistling shrieks of young macaques playing.

Langurs are the prominent species in the forest, and their lively antics often put them centre stage in the daily jungle drama. Langurs were clearly the inspiration for Rudyard Kipling when he wrote about the Bandar-Log (monkey people) in the *Jungle Book*. '*Bounding, crashing, whooping and yelling the Bandar-log swept along tree roads ...*' perfectly describes the behaviour of an over-excited all-male gang of langurs.

According to ancient legend, Hanuman, the monkey-god, helped Lord Rama fight the demon Ravana in Sri Lanka. The monkey armies then built the Bandhavgarh fort for Rama's brother Lakshman. The epic Ramayana tells that while in Sri Lanka, Hanuman stole a mango plant and brought it to India. Punished for his theft, Hanuman was set alight and in putting out the fire, blackened his hands and face. Possessing the longest tail of all monkeys the langur is aptly named – in the ancient Indian language Sanskrit, langulin means long tail.

Langurs are very social primates and in the forest are normally found feeding or resting near herds of chital – the deer and monkeys work together to keep watch for tiger or leopard. Eyes and ears, both on the ground and in the trees give twice the protection. Langurs feed mainly on leaves or flowers and sometimes offerings left at Hindu temples where the monkeys are considered sacred. These vegetarians need to supplement their diet with minerals and regularly visit established 'lick' sites at river-banks or waterholes, where large groups sit chewing at the salt rich earth.

Shortly after the rains end, the three boys – Raj, Sundar and Barra Larka – move away from their mother. There is still a strong bond between the brothers and for a while they stay together, successfully hunting in the area around Raj Behra meadow. For such young tigers, they have been extremely fortunate to be able to move to this high quality habitat with an abundance of prey. The siblings are in a transition stage and though self-sufficient, still have plenty of time for games and relaxing. Charger, the boys' grandfather, is still the dominant male in most of the Tala range and though he may currently tolerate the sub-adult males, the situation cannot continue for very long.

The boys are fast becoming adults with a need for a territory of their own. Although a tiger's territory size is affected by many factors, such as terrain, vegetation and water supply, the most significant factor is the availability of prey. To ensure a continued healthy prey population, it is estimated that, annually, tigers can kill about 10 per cent of the available prey, while still maintaining sustainable levels. As a rough example: if a tiger needs to eat 50 deer a year to survive, then for each tiger in an area, the deer population must number 500 animals.

Very hot and carrying an injury, it is easier for Charger to steal from inexperienced tigers, than to hunt himself. While Bachchi's boys dozed in the heat, blissfully unaware, Charger purloined their recent kill. However, when the boys realise what has happened they follow Charger's trail in the hope of winning back what should be theirs. The confrontation becomes a stand-off. In the past, Charger would have taken the youngsters on, and won, but now he is old and tired. The boys have not quite gained the confidence to challenge him, though they do make a half-hearted attempt. With every day the young males are growing in confidence and Charger's stature is declining. This may be his last stand – he is about fourteen or fifteen years old – a very impressive age.

For now Charger retains the kill. Drained by the morning's events and blistering heat, he is unable to summon up the energy to eat. Charger must find water, so he covers his prize with leaves to protect it from the forest's many scavengers. Indian monitor lizards, jackals, vultures and jungle crows will all help themselves to a free meal given the chance.

Too hot to eat. Charger heads to water, through the sal forest towards a large pool in the middle of Raj Behra meadow.

Charger may be past his prime, yet he still commands respect wherever he goes and is certainly able to intimidate unwanted wildlife photographers!

Sitting on long-since fallen tree, a lone langur relaxes in the golden, late-afternoon summer sun. A huge meadow with a year-round water supply, Raj Behra is home to many species. Storks, peacocks, rollers and vultures are regular visitors to the meadow. White-winged kites and crested serpent eagles hunt from above, while at ground level, the grasslands offer one of the best hopes in Bandhavgarh to glimpse the elusive jungle cat, an almost domestic-sized relative of the tiger.

And when the tiger is not in residence, there is a chance of spotting a leopard.

In the summer of 1999, Raj Behra meadow seemed a busy spot for tiger activity. Charger still uses the meadow, but it is his grandsons Raj (named because of the meadow) and Barra Larka that have set up permanent home here. A small stream runs through the grassland and it is there that Raj rests to cope with the high temperatures.

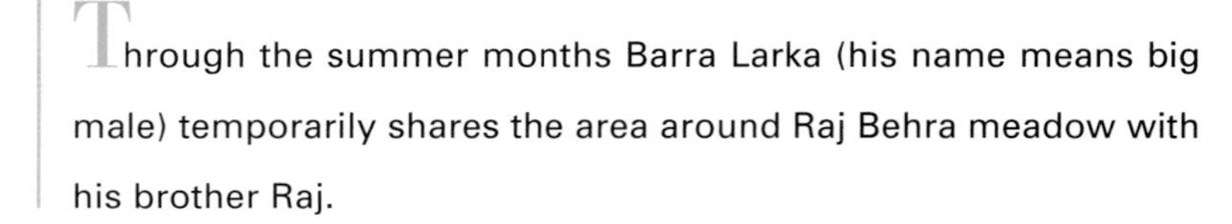

Through the summer months Barra Larka (his name means big male) temporarily shares the area around Raj Behra meadow with his brother Raj.

Sundar is spending less and less time with his brothers. He already has the urge to find a space of his own.

Summer is the time for peacocks to display, battle and seek a hen. Males often have favoured dusty arenas to perform to onlooking females. With such refined splendour they slowly spread their shimmering tale feathers. It is fantastic to watch these huge birds fly, then land in trees with such apparent grace. Peafowl are impressive birds in so many ways – their haunting 'kee-oww' calls provide one of the most memorable sounds of the jungle.

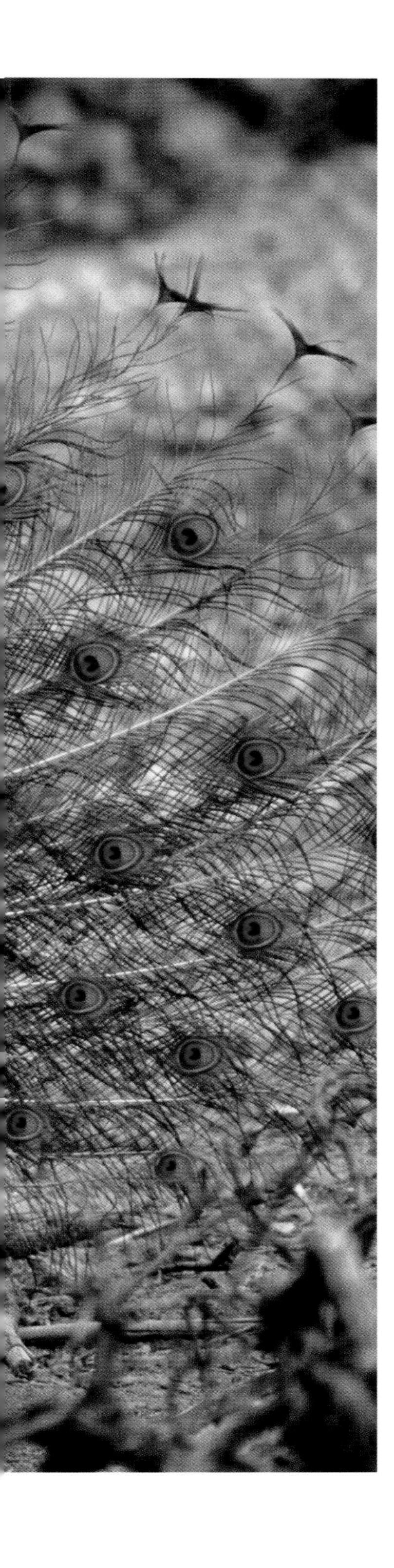

Amid the tired dry jungle there are some striking colours on display. Flashes of brilliance appear as Indian rollers make regular insect-catching flights. Their plumage has an infinite palette of hues, from ink blues and iridescent jades to dusky pinks – the colours change with every move of the roller's wings.

# The Brothers' Fort

2000-2001

With the dawn of the new millennium comes a new era in Bandhavgarh – Bachchi's boys are now the brothers of the fort. Old and frail, Charger cannot compete with the youngsters and is slowly losing ground – it will be his last season in charge. The brothers are gradually taking control of their grandfather's prime territory, equally splitting it between themselves.

Bathed in the early morning golden light, Sundar confidently rests in the dew-wet grasses in Chakradhara meadow – the grassland at the heart of the Tala range. To the west, Raj has established himself in the area around Raj Behra meadow and, having moved away from his brother, Barra Larka has gone east into the sandstone hills around Khirki.

And with the frosty mornings comes wonderful news from behind Badhaini hill – Bachchi has a second litter of tiny cubs.

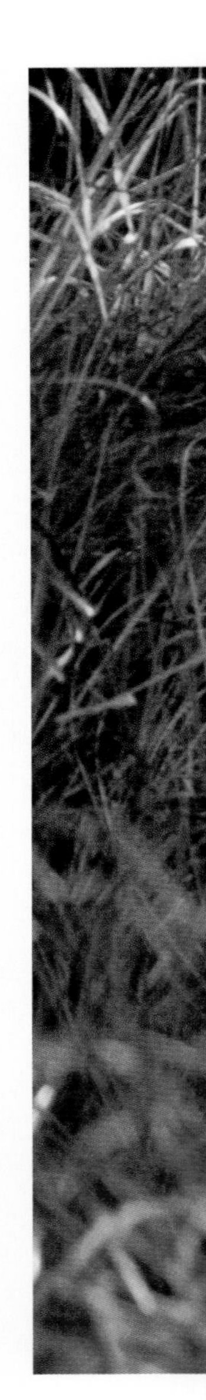

Nestled below the fort cliffs, Chakradhara meadow was once the site of a small village and farmers' fields. Now prime tiger habitat, the grasslands have been at the centre of Sita and Charger's empire for almost a decade. However, that era is over. Sundar, or B2 as he is also known, has his sights set on this territory – and there is little resistance coming from Charger. Old and unable to compete, Charger is keeping out of the way of his grandsons.

In early January, the peak of the Indian winter, the damp meadow is at its coldest and temperatures may almost dip to zero. There is even frost in wet areas of the grassland. Surrounded on three sides by hills, the early morning sun takes a little longer than normal to rise far enough to reach Sundar. Enjoying the warming golden light on his fur, the contrasting black and orange colour of his coat blends perfectly with that of the grasses. Seeking a quiet spot to sleep or privacy to eat undisturbed, the tall vegetation is perfect.

Sundar shares his new territory with a host of other wildlife. Some are on his menu, but most he will pay no attention to. Vine snakes, cobras, frogs, turtles and a multitude of spiders are the secret inhabitants of Chakradhara.

Bandhavgarh's meadows are, without doubt, at their most beautiful in early winter. The warm sun and huge swathes of winter-flowering grasses create a stunning display. Some of the grasses produce masses of white fluffy plumes making the meadows appear as if it has been snowing. This is the closest Tala's tigers will ever get to the real thing!

Enjoying the insect bounty, many species of bird make their home in the grassland. Vibrant green bee-eaters and electric-blue Indian rollers are the most eye-catching. More subtle are the creams, browns and greys of the bay-backed shrike, but they are just as beautiful. Through the cold months, the shrikes are a regular sight on the tallest of the grass stems, making occasional swooping darts to catch prey from the ground or in the air.

Deep in a thick stand of bamboo on the other side of the Tala range, Bachchi has hidden her two tiny cubs. The two sisters are about three-months old and will have only recently begun exploring with their mother. Locating the well-concealed cubs in the dense vegetation is made more difficult by the contrast of bright light and shade. One of the youngsters climbs onto a fallen tree, and appears right in the open. Unsure of what to do next, the cub walks up and down the log a few times and then leaps into the undergrowth in response to Bachchi calling her.

The brief glimpses of these tiny cubs were such a privilege because seeing them so young is very unusual. The cubs soon disappear into the jungle and are not seen again until well into the summer. Bachchi is an expert at keeping her cubs well hidden.

By the summer, Charger is partially blind and arthritic. Over the last six months Sundar has pushed him to the edge of his former territory. Thaudi meadow was Charger's last stronghold, but now Sundar hunts in the meadow and sleeps under the tree that was once the old male's favoured spot. Charger is thought to be fifteen or sixteen years old, an incredibly impressive age for a wild male tiger to reach.

During the monsoon, the old tiger's health fails and one day he was found resting near Dhamokha village. A weak tiger so close to human habitation was too dangerous, so the forest rangers had no choice but to usher him back into the jungle. Such a famous and loved tiger, the authorities took the unusual decision to build a compound for him in the forest. Provided with food and water he stayed in his jungle retirement home, sleeping most of the time, until he died on 29th September 2000. His body was cremated and the ashes scattered in the jungle. His last resting place is now permanently marked with a memorial stone. Every year thousands of tourists visit the memorial, pilgrims for the old King of the Jungle.

Sundar has gained the prime range and now takes on the habits of Charger. He has territory to patrol and may walk fifteen to twenty kilometres each day. With each purposeful stride he smells the air and looks or listens for signs of other tigers. As he travels, he will scrape-mark the ground or scent mark on trees or rock. These 'notices' have many purposes. They announce his presence to any females in the area – a typical male's territory would probably include the ranges of more than one tigress. The signs are also non-combat messages to assert his ownership of the area to other males. Sundar and his brothers are incredibly young to hold such good territory. If a slightly older or more experienced male arrived in this part of the forest, each of the brothers would have to work very hard to retain their new home.

Sita's daughters, Reshma and Pyari are now almost four years old and both have prime territories. Reshma lives in the rocky area around Banbehi in the east of the Tala range - where Barra Larka has recently moved to. Though this habitat is not as rich as some parts for prey species, it does have numerous waterholes and streams; also, importantly, the terrain of hills and ravines provides plenty of hidden locations where Reshma could give birth and hide the cubs. She has not yet had cubs, but is certainly now of the right age and will be paying attention to Barra Larka's arrival in the area.

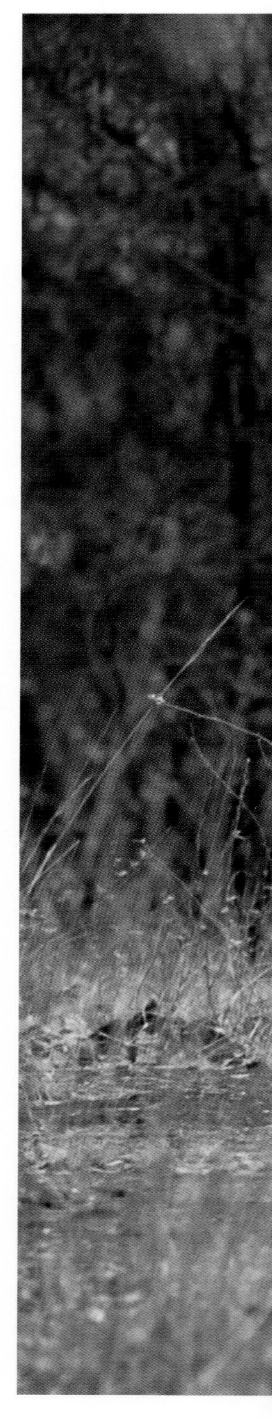

Barra Larka has three primary tasks to undertake in his newly independent life. Firstly he must protect and hold his new territory, which involves him walking through as much of his territory as frequently as possible, to assert his ownership. Though they continue to evolve, he already has established routes to walk and favoured places to scent mark or trees to scratch with his claws. Barra Larka will vocalize, to communicate with the other tigers, but in the forests of Bandhavgarh it is uncommon to hear tigers growling.

The second job is to find food – Barra Larka has a much hillier and prey-poor habitat than Sundar. With few lush meadows, most of the sparse vegetation is to be found on the rocky sandstone areas. Sambar deer are the most abundant grazer in this area and will form the bulk of Barra Larka's diet. He will also hunt the smaller and more secretive muntjac, or this female chowsingha (four-horned antelope - as seen above). And finally Barra Larka must find a mate!

Quickly stepping off the road into the undergrowth, Barra Larka disappears for a brief moment; the monsoon will come soon and this may be the last sighting of him until after the rains. However, when he reappears there is tremendous excitement.

Lying on the forest floor, Reshma observes Barra Larka intently as, cautiously, he steps nearer. Jungle life seems to stand still as the two tigers reach forward and touch noses, almost a tiger kiss. For a few seconds they smell each other, then, with great care, Barra Larka settles down next to Reshma. The meeting is a highly charged one. Though the tigers appear comfortable, there is definitely an unsettled tension.

Tigers are solitary animals and there should be only one explanation for this meeting – but the pair do not mate. For almost an hour they lie together, frequently changing positions. Each time one tiger moves, there is a noticeable awareness of the other. Without an obvious prompt, Barra Larka gets up and slowly saunters off down a slope into the jungle. He repeatedly calls to Reshma, though she does not follow.

It is a few weeks later before an explanation for this encounter is found. While on a regular patrol, one of the elephant rangers discovers four tiny cubs concealed in a nearby ravine. The gestation period for tigers is approximately 103-108 days (or three and a half months) so Reshma must have been almost three months pregnant at the time of the meeting. Barra Larka was almost certainly the cubs' father.

By March 2001 Barra Larka is well established as the territorial male; whether it is due to his wary nature or the inaccessible terrain, he is seldom seen. Barra Larka and Rershma's four cubs are growing fast; the three males and one female have inherited a strong character trait of their parents and remain incredibly elusive. Their mother's name means 'silken' because of the way Reshma seem to glide silently through the forest.

With night-time fast approaching and a sizeable stretch of sandstone and deep ravines separating him from human onlookers, Barra Larka is very relaxed.

# Tiger Jungle

Reshma's sister, Pyari, also has cubs born at the same time – one male and one female. Pyari and her two cubs, Babu and Narangi, live in and around Chakradhara meadow, a prime habitat. She has an abundance of prey and water in her territory and the lower edges of the fort cliff provide superb locations for raising or hiding young cubs. The hill slopes are peppered with caves, both man-made and natural – the jungle even conceals secret temples and statues. The two cubs spent much of their early months playing in and around the overgrown Barra Sheshshaiyya temple.

One afternoon when they were about eight months old, the siblings appeared on a jungle-clad boulder high up on the hillside. Well out of harm's way, they were incredibly relaxed as they surveyed the forest activities. Waiting for Pyari to return from hunting, the cubs played, sunbathed and stretched on the large rock. Narangi sits in the open while her brother enjoys the cover provided by the tree foliage.

A few hours later the cubs left the hillside – their mother was calling from Chakradhara meadow below. Earlier, Pyari had made a kill nearby and during the quiet afforded by the intense midday sun, she dragged the heavy carcass the considerable distance into the meadow.

Padding through the bamboo leaf litter Babu explores th
jungle while Sundar rests close by. For another year or so, fathe
and son will happily co-exist, but there may come a time whe
Babu becomes the challenger to Sundar.

Territorial males will tolerate young or transient male tigers unt
they are about 30 months old – the age when they becom
sexually mature.

Anuthi left her mother during the monsoon and due to the lack of available space was forced to move out of the Tala range and in to Khitauli. Tracking or studying her in her new area is difficult, but forest rangers have since seen her with at least one litter of cubs. Sundar regularly visited the area and was almost certainly the father of Anuthi's first litter.

4

# Tiger, Tigers Burning Bright

2002-2003

In the forests ... of Tala, it is a magnificent time for the tigers. Sundar confidently moves through the parched grassland terrain, sniffing and scent-marking as he walks. Five years-old and well established in the heart of Bandhavgarh, Sundar is in his prime. The other brothers are settled in their ranges – Barra Larka in the hills, enjoying the privacy this inaccessible terrain provides. Although Raj is rarely seen, his mate, the tiny tigress known as Kokila, is raising three cubs near Raj Behra meadow.

Inhabiting the jungle, there are four tigresses with cubs of widely varying ages. Slightly older than Kokila's cubs, Babu and Narangi are nearing dispersal age and these cliff-raised cubs have taken to spending time on top of the fort hill and scaring the temple priest. Babu has also been observed stealing the milk from the mouths of babes! Reshma's cubs are also nearing independence, so she and Pyari may soon have a second litter.

Below Badhaini Hill, Bachchi has a third litter of much younger cubs, which in her true style are kept well hidden. Observed mating with Bachchi in June 2001, Sundar is the father of these cubs.

On a forested hillside behind Raj Behra meadow, Kokila's three cubs lie secreted among the undergrowth. It is mid-morning and the only activity they indulge in is the occasional yawn or a spot of grooming. The male and two females are about 18 months-old and like Tala's other teenage cubs, they spend most of their day sleeping or waiting for Kokila to return with food. Fortunately she has a sizeable range with plenty of prey, so the youngsters do not go hungry. Often with three cubs, a special bond develops between two of the siblings. The male (K1) and his sister (K2) had that relationship and always moved together. As if attached by elastic, whenever one got up, the other quickly followed. K3 naturally became the most independent of the trio.

Raj and Kokila are seldom seen, partly due to their secretive behaviour, but mostly because of their expansive territory which extends through Bagh De Laka into the adjoining Kalwah and Magdhi ranges. Much of the knowledge of their movements comes from identifying pugmarks along the sandy tracks.

During the monsoon, Kokila and Raj moved away from Raj Behra - probably into the adjoining range to assist their newly independent daughters to establish a territory. When cubs reach dispersal age, young females may inherit a portion of their mother's range, or gain it all if their mother relocates. Young males do not have it so easy and must leave the natal area. Territorial tigers will tolerate these young transient males for a few months, but eventually they will be forced to move on, until they find vacant land elsewhere.

Assuming there is a healthy tiger population, transient males will start in quite poor terrain on the edge of the forest and over time they will attempt to move into better habitat. Where poaching occurs, unnatural gaps appear and this may allow relatively inexperienced tigers to gain a good area. Bachchi's boys were extremely fortunate, because of their grandfather's age, to take over prime territory when so young. Almost certainly due to poaching in the adjoining ranges, no older males were present to challenge Charger.

From where he came no one knows, but there is a new middle-aged male in Raj's old range. Searching for a vacant prime habitat and a receptive tigress, Shaki has arrived in Raj Behra. He has a very distinctive facial appearance, very unlike Bandhavgarh's other males, which suggests that he is unrelated and has travelled from outside the immediate area.

Shaki is interested in Juhi (K2) and Chameli (K3). The tigresses have stayed in the area, but they are very young and probably on the borderline of being sexually mature. It was widely thought that females could not conceive until three or four years old – the sisters are only two and a half. However, studies in Bandhavgarh, by Kay and Satyendra Tiwari, show that tigresses begin mating as early as 30 months. Pyari first mated with Sundar at this age, though because he was younger, their early pairings were not successful.

Juhi seems relaxed, but her new suitor is unwanted.

Lying on the forest floor together, the two tigers appear to be at ease, however there is definite tension. Juhi tries to move away from Shaki, but he is very persistent and follows her into the dry nala. An abundant feature in Bandhavgarh, these sandy channels are the routes of seasonal rivers that only flow during the monsoon. Throughout the rest of the year these waterless highways form natural walkways for tigers.

Besieged, Juhi erupts. Mustering all her aggression she turns towards Shaki and unleashes a phenomenal roar. The ferocity and sound is incredible. It is such an unusual occurrence, that the noise silences the jungle. Shaki backs away for now, but he will try again.

It is not known whether they eventually mate successfully, because not long after this encounter, Juhi moves away from the area. However, his patience paid off with her sister Chameli, when in early 2004 they mated successfully.

Far, far away on Badhaini hill two leopards crawl like tiny insects along the distant escarpment. Normally elusive, the leopards are quite relaxed and watch the forest events from their spectacular vantage point.

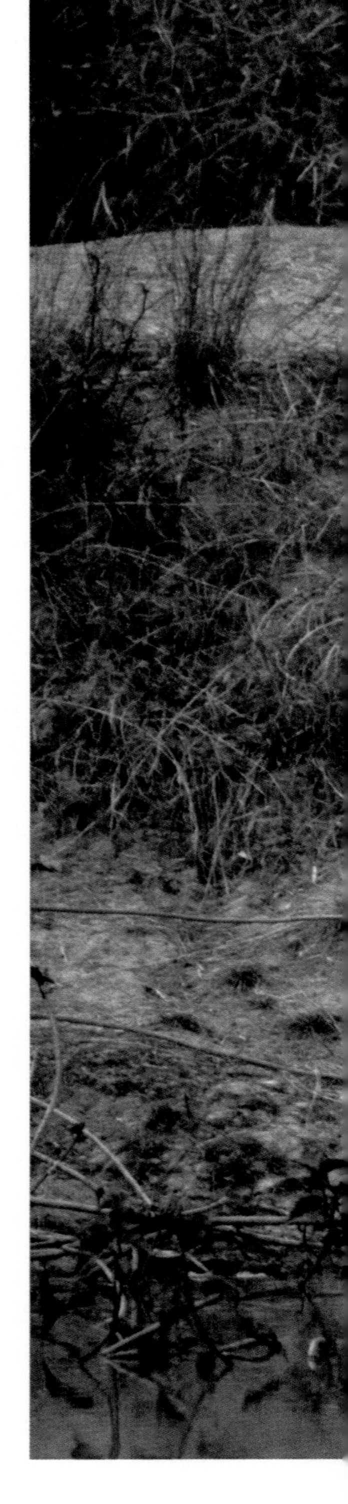

Reshma is just visible under a low bush near Banbehi waterhole; her cubs are perfectly concealed nearby. In April 2002, both Reshma and her sister gave birth to a second litter of cubs. It was thought that in an unexpected territorial crossover, Sundar fathered Reshma's three male cubs, and Pyari's four are Barra Larka's. However, subsequent events suggest that pairing was the same as with the first litters.

Whereas Reshma's first litter have all dispersed, heading out of the Tala range, Pyari's cubs, Babu and Narangi, are more reluctant to leave. At two and a half years old, they still live on the edge of Pyari's territory. There were some extraordinary scenes in November, when the adult Babu was seen suckling from his mother, alongside her new litter of seven-month old cubs.

Sundar will eventually have something to say about his son's unwillingness to leave home completely. However, for the moment Babu is helped by the fact that Sundar is absent for long periods. He has a particular interest in Anuthi and is spending much of his time in her territory and away from Chakradhara meadow.

Emerging from a cooling bath in the Julwani waterhole, Pyari heads back into the jungle to her nearby cubs.

Pyari has three female cubs and one male to provide for, so there is a constant pressure on her to hunt successfully. With five mouths to feed, and occasionally Sundar as well, she must make regular kills – by March 2003 Babu had left the Tala range. Empty-bellied, Pyari stalks a sambar, her focus entirely on the prey. Slowly she approaches her quarry using the undergrowth as cover; tigers rely on stealth and ambush to get close enough for a final brief charge. The careful approach takes almost ten minutes, but then some slight sound or smell alerts the sambar to her presence. Instantly the deer begins to thump its foreleg on the ground and let out the distinctively resonant barking alarm call. Pyari has failed this time, but judging by her cubs' healthy growth, her success rate must be high.

On a beautiful morning in March 2003, all four of Pyari's cubs entertain themselves on a small sandstone hillock. The very nature of the jungle habitat means it is unusual to see all the cubs together. The young tigers are ten months old – an age where they are incredibly inquisitive and playful. One of the females, known as Lakshmi, climbs down the rocks and starts tapping a loose chunk of stone with her paw – sniffing and pushing it with her nose she sends it scuttling down the slope.

Indrani watches Lakshmi, playing on the rocks. The third sister, Durga, appears at the top of a nearby rock to gain a better view.

Dozing in the dense grass in Chakradhara meadow, Lakshmi rests her head on her mother. Having recently eaten, Pyari and the cubs are resting with full bellies. More for comfort and bond than food, the cubs regularly suckle from their mother.

Through small gaps in the undergrowth there are tantalising glimpses of Lakshmi and her siblings.

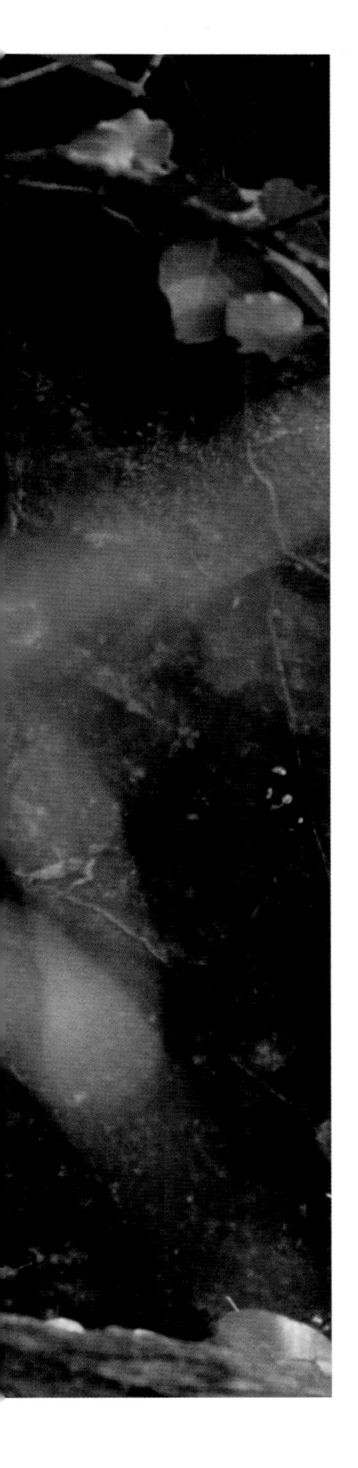

There are few finer moments in wildlife photography than discovering four young cubs together with their mother, lying on a sun-drenched ledge. Over the course of the next hour the late afternoon light warms and the scene becomes even more enchanting. Pyari keeps watch from the back of the ledge, partly hidden behind a small Sal tree, while two of the sisters lie in the open. Pyari and her daughters have recently finished eating. The male cub is still feeding on the remains of a sambar deer in the dense bamboo below.

Initially hidden in shady spot, Lakshmi wanders along the edge of the ledge to join her sisters.

Into the light, Lakshmi settles down in the golden late-afternoon sun in front of her sister, Indrani.

Learning from their mother, the cubs are already attempting to hunt. Shortly before dusk, Lakshmi and Durga have spotted a female Chital and her fawn on the edge of Chakradhara meadow. Unfortunately for the cubs, the deer has also spotted them.

Undaunted, the young tigresses continue to approach, using the tall grass as cover. The failing light works in their favour; at times the cubs are barely visible.

The mother and fawn have long gone, so the cubs continue the evening exercise as an educational game. With one final dash they ambush an imaginary deer. The cubs are still a year away from needing to hunt for themselves. Though their skills may require a little refinement the sisters are well on their way to hunting successfully.

Summer has come early to Bandhavgarh; although it is only early March, the days are already becoming hot. Sundar avoids much activity in the daytime; he seeks out a place to bathe or one of his many caves to sleep in. On the edge of a dry river bed this ancient cave was probably carved more than one thousand years ago.

Though warmer than usual, the jungle is looking particularly colourful with many trees in flower. The resplendent blooms of the palash tree, or flame of the forest as it is also known, are enjoyed by numerous species of birds. Parakeets, babblers, mynas and black-rumped flamebacks (a type of woodpecker), all take advantage of this natural bounty.

## A Forest on Fire

2003

In early summer 2003, Bandhavgarh is on fire with the rich coral-orange blooms flickering from the flame of the forest. The mass flowering is especially impressive this year, the best for almost a decade. Vistas across the forest canopy reveal great swathes of orange. But not just the palash, for the silk cotton tree with its brilliant yellow flowers is also unusually resplendent. Sparks of deep red adorn the bizarre bonfire tree that appears to grow upside-down.

However, the forest is also ablaze with news of dead tigers, poaching is raging through the jungle, destroying lives and years of growth. The year is a very sad one for Bandhavgarh and for all Indian wildlife. In separate tragic incidents the jungle loses two of its best-known characters and further afield Chinese authorities made a shocking seizure of many hundreds of wildlife skins, which were on their way from India to Tibet.

On a forest track in Bandhavgarh, six tourist jeeps wait quietly, eager for a glimpse of a tiger. Eventually, a tigress appears in the distance walking slowly towards them. None of the occupants were ready for what they would witness next. On 29th March, in a freak incident, the approaching tigress jumped inside one of the jeeps and began to attack two French tourists. A local lodge owner, who was sitting in a neighbouring jeep, intervened and bravely tried to fight off the tigress with a small log. After almost five minutes the tigress backed-off and the three injured people were rushed to hospital. Fortunately, their injuries were not as serious as would have been expected.

This type of attack is unheard of; at first there seemed to be no motive. However, witnesses said that the tigress was very weak, staggering and missing three canines. The single puncture wounds and limited injuries sustained by those attacked supports this.

Earlier in the morning, the same tigress damaged a forest guard post situated along the main road. Nearby, on a roadside boulder, deep scratch marks had been found, together with blood, tiger hair and fragments of tooth. It is probable that the tigress had been hit by a vehicle while crossing the road, causing the loss of many teeth and other unknown injuries. Upon encountering the jeeps, the frightened tigress acted aggressively in self-defence. Initially, the injured animal was thought to be a tigress new to the area. However this was Bachchi's territory and in the weeks following the incident she was not seen. Subsequent studies of video footage, shot at the time by tourists, show that it was indeed Bachchi. She must have been very badly hurt to make this uncharacteristic attack in the first place. And then to not have the strength to carry it through.

The reasons behind the attack may never be completely understood, but some issues need to be addressed. A reduction of traffic on the main road, which bisects this part of the forest, would be a good start. Ironically, the following month, work began on widening the road.

Many people will have read about Bachchi and her family and some have been lucky enough to see her. An elusive tigress and a superb mother of eight cubs, she will be greatly missed; though the forest remains alive with her legacy of young tigers. Bachchi left three eighteen-month-old cubs – this photograph was taken less than a week before she died. The cubs were close to dispersal age, but given the chance, they would probably have spent another two or three months with their mother.

Left to fend for themselves prematurely, the young tigers must use all the skills that Bachchi has taught them. For a while the cubs stay in the same area, as if awaiting their mother's return. During the next month it becomes clear whether the cubs are able to survive without their mother.

Bachchi's trio have been able to make a number of kills themselves. However, a few months after their mother dies, one of the cubs sadly succumbs to a seemingly innocent jungle danger. Porcupines are not especially dangerous animals, but inquisitive young tigers can easily injure themselves on the quills. Without an experienced mother to guide him, B6 dies from getting a quill stuck in its mouth.

Over the next few months B8, Bachchi's adolescent female stays in the immediate area, though long-term she is probably too young to hold the territory. It is likely that soon she will move completely into the adjoining Khitauli range. Her brother, B7, is also still in the area, but before long he too will have to relocate. He has a slightly injured paw causing it to drag a little, making identification of his pugmark easy. His disability may hinder him in adult life – one that began so abruptly. Bachchi's two remaining cubs were last seen, during the monsoon.

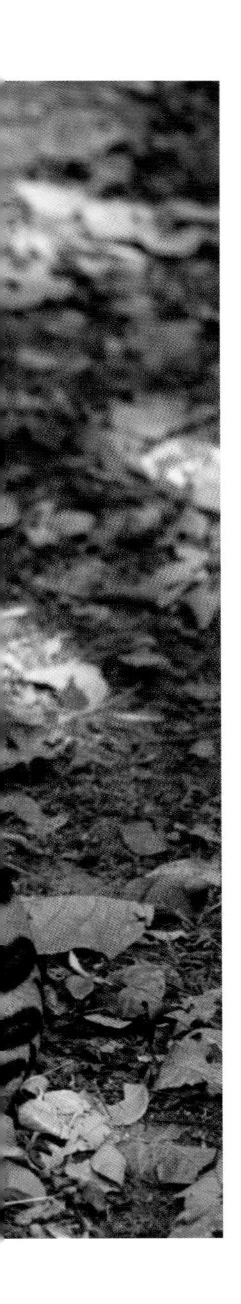

In early October, 2003 Chinese authorities made a shocking seizure of wildlife skins that were on their way to Tibet. Subsequently confirmed to have come from India, the skins included 31 Tigers, 581 Leopards and 778 Otters. This one haul of pelts probably represents about two per cent of India's entire tiger population and an unimaginable dent on leopard and otter numbers.

Only a few weeks later there is a distressing discovery in Bandhavgarh – the skinned body of a large male tiger is discovered. The electrocuted body was found in the area of forest that is the range of Barra Larka. Praying that it isn't him, our hopes drop as the weeks go by without any sightings. Sadly, he hasn't been seen since and it is now clear that Barra Larka died at the hands of poachers on 29th October.

The poachers are either locals who know the forest intimately, or nomadic tribal people for whom it is a hereditary profession, moving from forest to forest, hunting and trading. Those who undertake the actual killing are only the start of a chain. Via middlemen, the many poachers pass their wares to a few key traders who are the masterminds behind the trafficking. Mostly based in the big cities, these powerful traders, like drug mafia, are the most important link in the chain and often avoid punishment. Although arresting a poacher or a retailer is good news, there are unfortunately many more to replace them. Control of the wildlife trade, however, is limited to a few powerful people; remove these links and the chain will break.

Poaching is big business and needs tough, thorough action to combat it. Many non-governmental organisations and enforcement bodies working hard in the field are continually disappointed by the weakness of fines or jail sentences and more often, the complete failure to even bring about a prosecution. One major reason for the collapse of past prosecution cases is the confusion, rivalry and bureaucracy between the different departments investigating the incident. Despite recommendations from an Indian Governmental Committee in 1994, it took until February 2002 for India to agree to the setting up of national wildlife crime unit responsible for investigating and eliminating the wildlife trade. Despite all the agreements, the unit is still not in operation in 2007. Significant political support from the highest level is needed to ensure that the unit is set up immediately, and with sufficient resources, including personnel.

A grim twist to the illegal tiger trade is the concern that China wishes to start trading in captive bred tigers. In China, more than 4000 tigers exist in huge farms/zoos, of which the largest establishment in Guliin has 1400 tigers. With freezers full of frozen tigers, this is big business and they want to be allowed to sell them for meat, medicine and skins. Amazingly, the owner of the tiger farm at Gullin was a member of the Chinese Government delegation to the recent International Tiger Symposium in Kathmandu, Nepal. He argues that farming tigers will end poaching in the wild. Aside from the horrific conditions and lives of these animals, conservationists are convinced that it would create an open season for trade in wild tigers. Illegal wild tiger products would be easily laundered among the farmed animal trade.

Poaching for the illegal wildlife trade is a serious threat to the survival of tigers, leopards and many other Indian species. If the trade is allowed to continue at the current alarming level, then it will only be a matter of time before tiger populations are wiped out, in all but a few areas. It would be a very sad indictment on us, if India's once rich wilderness were confined to a few artificial arks.

Conservation efforts are not purely about preventing the tiger from becoming extinct, but rather protecting the forest habitat as a whole. Securing long-term protection for significant areas of habitat is not just saving a place for the tiger; it will ensure the survival of thousands of other wildlife species and help to provide a healthy environment for the people of India to live in.

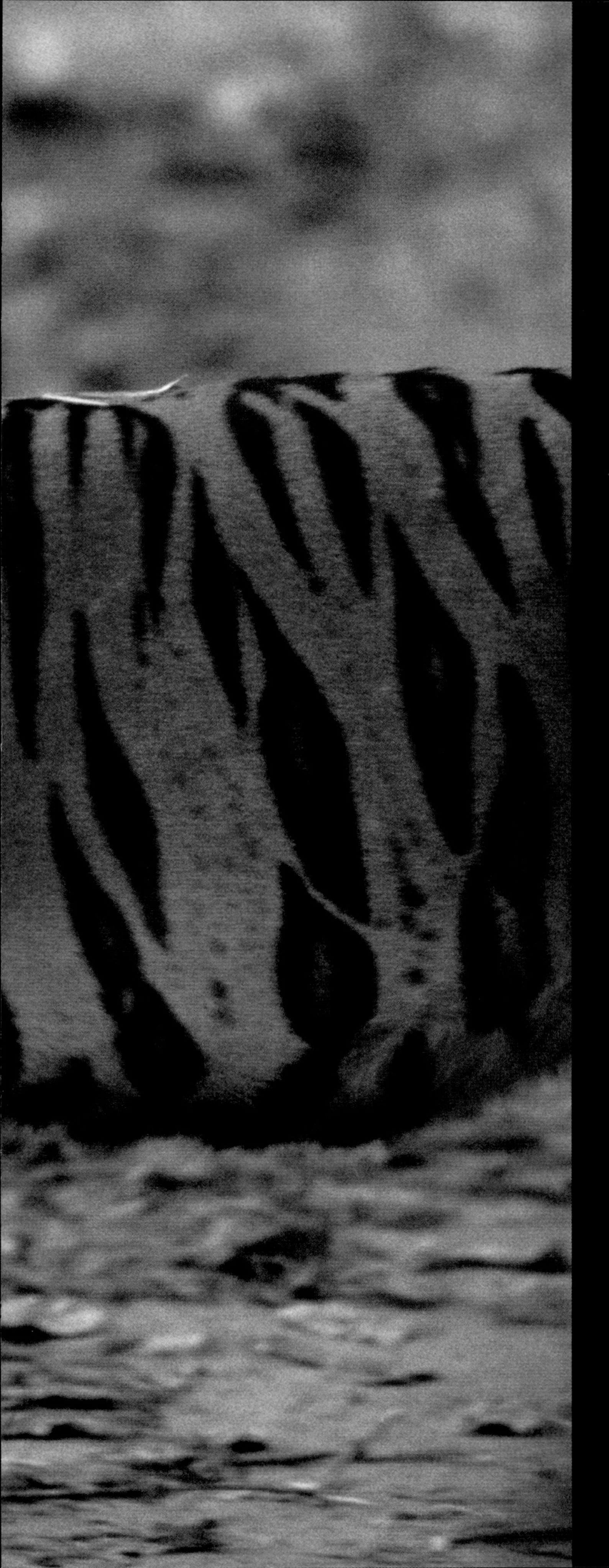

# King of the Jungle

## 2004-2005

As the dust settles after a traumatic year, Sundar emerges unscathed and enhanced. King of the Jungle, he has absorbed Barra Larka's range into his and so has become the territorial male in both Pyari and Reshma's area. Sundar shares the forests with a diversity of predators and battling species, but for him it seems that the only real rival is Shaki. Until Babu returns. In early 2005, Sundar's first male cub, wanders back into the central range after an absence of more than two years. He must have been living in the depths of one of the other ranges or in the buffer-zone, as there had been no reported sightings of him. Babu's reappearance does gives great hope for the future of all young tigers that must leave the relative protection of Bandhavgarh's core area.

Pyari's daughters were the lucky ones and did not have to the leave area they knew well. They moved into the vacant area of jungle left by Bachchi and by taking a good slice off the edge of their mother's territory, the sisters have carved out three small adjoining ranges.

The territories of Tala's newly adult tigresses, Lakshmi and Indrani, fall within Sundar's area and he wastes no time in trying to mate. As is normal with tigers, Sundar mates with Lakshmi regularly over a three-day period. However, nothing comes of this pairing; Lakshmi is two and half years old and may not yet be sexually mature. It is almost a year later that she has cubs – one male and one female. Born in late 2005 it is not clear who the father was, as Babu, Sundar's son and challenger, also mated with Lakshmi. Encroaching on Sundar's territory, Babu is living on the edge of Tala around Thaudi meadow. Early in the year he avoids conflict with his father by regularly crossing out of the Tala range into Khitauli. However through 2005 the two males gradually come into conflict more.

Sundar's other rival, Shaki, is the father of Durga's two cubs. One male and one female, they were born in May 2005. It is thought that Indrani also mated with Shaki, producing two cubs. However, apart from some early sightings they have never been seen and it is now clear that they did not survive.

In early summer 2004, both Pyari and Reshma gave birth to their third litters of cubs. Reshma has another large litter, one male and three females, which she keeps well hidden in the hills near Banbehi waterhole.

Sadly, Pyari's third litter did not survive for more than few weeks – it is not known why. Just before Christmas there is some very exciting news – the very brief first sighting of Pyari with tiny cubs, thought to be about five weeks old.

Reshma's seven-month old cubs rest and explore the varied terrain of their mother's territory.

All tigers have white spots on their ears and it is widely thought that these markings are to help cubs and adults locate each other in the dense vegetation. Their coats offer perfect camouflage merging well with the long dry grass – whereas the white flashes act like a beacon.

In the haze of an enchanted winter dawn, a pair of golden jackal trot through Sehra meadow. The jackals are on an early morning search for food – be it rodents, birds, reptiles or even forest fruits. They do not scavenge as much as is commonly thought, but will certainly take advantage of unattended kills made by larger carnivores.

Jackals normally live in pairs or small family groups and their presence is most noticeable when darkness descends on Bandhavgarh. The jungle and surrounding areas come alive with their howling harmony. Listening from villages and camps on the edge of the forest, the calls can be heard from every direction, as different jackal groups communicate with each other.

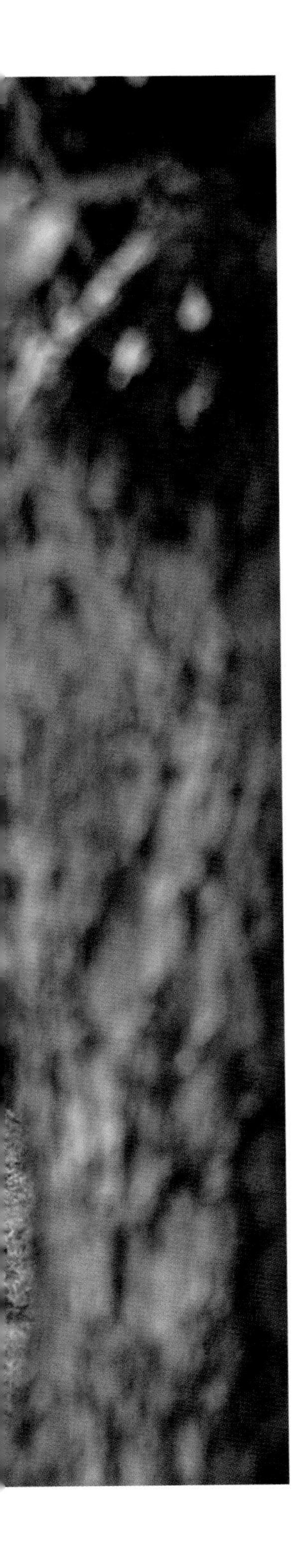

On sentry duty, a dhole keeps look-out while its mate feeds on a chital. With military meticulousness, the pair of Indian wild dog switch duties as they feed and keep watch. Observing them feed, offers an unique insight in to the incredible social skills of an animal that normally lives and hunts in large packs. With no visible or audible communication, the wild dogs seem to instinctively know when it is time to swap roles, which they do every couple of minutes. Feeding on the edge of Bathan meadow, the dhole must stay alert – tigers would kill them given the chance.

Indian wild dogs were once heavily persecuted by man, but thankfully today they are very welcome in the jungle. Their hunting techniques mean they tend to seek out injured or old prey and so help to keep the deer populations healthy. Dhole are a rare sighting in the Tala range due to the high density of tigers. However, in the other ranges where tigers are less concentrated, there are packs of up to twelve dogs.

Amazingly, at the end of 2005, Bandhavgarh has six adult tigresses and all bar Indrani, have cubs. Pyari, the older tigress in the central Chakradhara meadow, has four cubs that are now a year old. Growing well, the two male and two female cubs certainly benefit from their mother's hunting expertise. Lying in thick bamboo behind a dry river, Pyari rests while her cubs feed on a wild boar. She is an incredibly beautiful tigress, who has inherited the striking looks of her famous mother, Sita.

A lone rhesus macaque sits in a tree on the edge of Bathan meadow; the Bandhavgarh fort plateau forms the backdrop.

In the rich late afternoon light, a flock of Indian peafowl forage in a forest clearing

Chameli's boys were born on March 2004, following her mating with Shaki. These two cubs are the fifth related generation since the times of Charger and Sita. Bachchi was their great grandmother and Raj their grandfather.

In late 2005, Chameli's cubs are almost adult – the two, eighteen-month old brothers are already exploring the jungle independently. They live in the area where their grandfather had his territory. On many occasions in December 2005, the brothers were seen resting near, or crossing Raj Behra meadow. Watching the two young tigers in the grassland evokes strong memories of the summer of 1999 when a similarly aged Raj and his brother Barra Larka rarely seemed to leave the meadow.

Chameli's boys may stay in the area around Raj Behra for a little while, but eventually they will have to move on. Soon, their father, Shaki, will not tolerate them. In May 2006, Chameli had a second litter of three cubs.

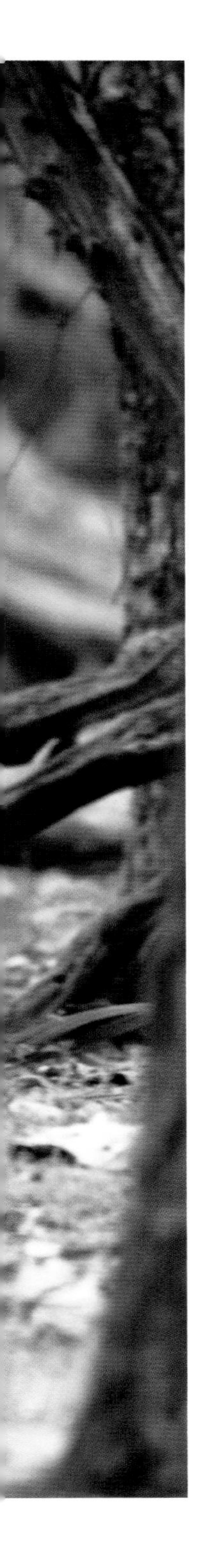

Sundar is still the king of Bandhavgarh, spending much of the time roaming his large territory. Due to his expansive range, he is increasingly difficult to find – sightings are a real privilege. Confidently moving through the sal forest, Sundar sniffs and scent marks the trees. However, although Sundar still controls most of the Tala range, Shaki is slowly stealing his territory.

The news of Sundar's son, Babu the challenger, is not so good. Through 2005, Babu had been seen regularly and was expected to carve his own territory in the north-western edge of the Tala range, in the same area as Indrani. However, in November Babu got caught in a poacher's snare and needed to be tranquilized by a local vet so that it could be removed. Only a few weeks later he was seen near Thaudi meadow with some nasty facial injuries. He may have fought with Shaki or Sundar; alternatively the wounds may have come from another bad encounter with humans. Sadly, on 9th December he was found dead. It is thought that he had died from his injuries.

# Prospect Tiger

## 2006-2007

From the fort cliffs the jungle view is simply breathtaking, however the prospect for its inhabitants is not so clear. Poaching and environmental pressures are placing a huge strain on India's wildlife. Nesting on a rocky outcrop just below the Maharaja's Seat, this long-billed vulture is a prime example of why the protected forests are so important. Populations of three vulture species, the long-billed, slender-billed and white-rumped, have plunged so dramatically that they are now listed as 'critically endangered' – even more at risk of extinction than the Bengal tiger.

However, tiger numbers have also reached an all time low. New reports suggest that the worldwide tiger population may be as low as 3500 animals, with less than 2000 living in India. One reserve has lost all its tigers and other well-known reserves have been hit by poaching. Project Tiger, the governmental body responsible for tiger conservation, has been unable to stop the decline.

Against all odds, Bandhavgarh is surviving well. The Tala range has its highest density of tigers ever. In 2007 there are seven adult females, six have cubs of varying ages, two adult males – Shaki and Sundar, and a number of young transient tigers. For Lakshmi's cubs, in particular, there is a dramatic start to the year.

*Photograph S.K. Tiwari*

Soaring on superb thermals, Bandhavgarh's vultures have a stunning view of the forest. As the day warms, they take to the air on warm up-currents to gain altitude, often circling at great height for hours. However, this beautiful scene does not show the true picture. Across India the vulture population has plummeted over the last decade and in many areas, species have been wiped out. For years, the decline was thought to be caused by a virus, however it is now clear that the crash is due to the toxicity of diclofenac, a livestock drug. These feathered 'cleaners' of dead animals, have paid a high price for feeding on cattle carcasses containing lethal levels of the drug. From the most abundant bird of prey in the world, populations of the white-rumped vulture have fallen by a staggering 99 per cent over the past decade – the Bandhavgarh population now numbers just a few birds – mostly seen around Tala village.

Luckily the long-billed vulture has not been hit to such an extreme in the park. The cliffs of the fort and neighbouring Badhaini hill still support reduced populations. Against the national trend, in April 2007, there were wonderful scenes of about 70 long bills feeding on a deer carcass in one of the central meadows. Probably because vultures within the national park feed on more wild animals than domestic cattle, they have not suffered the side effects of the drug as badly. Bandhavgarh is also home to red-headed vultures and the distinctively white Egyptian vulture – sadly both species are now showing signs of population decline – also attributed to diclofenac.

Thankfully the manufacture and import of the veterinary diclofenac has recently been banned in India, though it will take some time for any stockpiles of the drug to be used. An urgent international campaign has been launched by conservation organisations to prevent the unthinkable – skies devoid of vultures circling on thermals.

# Tiger Jungle

Play fighting in a forest pool, Lakshmi's cubs enjoy the cool water, blissfully unaware of the attention their activities are drawing. Earlier in 2007, while their mother was away, the two cubs had a tragic encounter near the forest boundary at Khitkiya. Illegally collecting firewood inside Bandhavgarh, four women from a local village came face-to-face with the fourteen-month old siblings. Naturally the women ran, but so did the cubs and sadly they caught one of the villagers. Aware that they were not supposed to be in the forest, the three women who escaped, delayed raising the alarm. When forest rangers eventually reached the scene late in the afternoon, much of the body had been eaten.

The cubs were only behaving naturally, but out of panic, there were plans set in motion to capture them and send the family to the nearby zoo in Bhopal. The cages and tranquilizers were ready, but thankfully a number of local people protested. Perhaps surpisingly, even local villagers handed in a 50-signature letter to the authorities stating that they did not want the tigers moved. They said the woman was only killed because she entered the tiger's territory. With such extraordinary support his impulsive scheme was dropped. Last year a tigress killed a person within the National Park, in the neighbouring Panpatha range. She was caught, together with her cubs, and sent to the same zoo, where today her pathetic form languishes in a small enclosure.

These terrible incidents highlight the daily conflict between man and wildlife. In India, where many people live in poverty, it may be argued that the protection of tiger habitats is a low priority. However, for many reasons, the survival of India's forests is just as important for the local people who live around it as it is to wildlife. These forests literally breathe life into India, improving air quality, providing clean water for the local people and bringing significant income and employment through forestry and tourism.

In India, millions of people in rural communities depend upon the vital watershed forests, which are protected within wildlife reserves.



Whether through mass clearance by timber mafia, indiscriminate mining and the resulting pollution, or slow denudation by villagers collecting firewood, bamboo or grazing cattle, India's forests and tiger habitats are under severe pressure. Once the forests could support normal harvesting, where fresh growth quickly filled the gaps; now India's wilderness is shrinking fast.

For the good of both local people and wildlife, the amount of protected and healthy wild habitat needs to be increased. However, this does not need to mean taking land away from people. There are large areas of degraded forest that are uninhabited by tigers because the vegetation is severely damaged and prey hunted to extremely low levels. A recent report stated that less than 40 per cent of existing forests in Central India can support tigers. So it is not a case of needing to create new habitat to save the tiger, just the need to properly protect the forests that already exist.

One of the saddest things about the villager's death is that it could so easily have been avoided. There are inexpensive fuel alternatives to wood such as simple solar cookers. If the large sums of money available to save the tiger were spent wisely, on a local level, so much could change. What is lacking is political will and a clearly thought out plan.

Ultimately, saving the tiger cannot be successful without an active community involvement. Conservation and the requirements of local people are integrated issues; if one is given priority there will be problems or a backlash from the other. India's remaining wilderness areas have an immeasurable value to the human population, be it environmental, social, aesthetic or financial. With tighter controls on industrial and commercial exploits, and a cohesive management plan, there is a chance that a balance can be found between man and nature.

*Photographs S.K. Tiwari*

Photograph S.K. Tiwari

**Sundar** - Ten years old

**Sundar** - Eight years old

**Sundar** - Five years old

For more than ten years Sundar has traversed the Tala range. His successful journey has taken him from one of three young brothers, to the dominant territorial male of the Tala range and father of more than 20 cubs. Living in the prime jungles at the heart of Bandhavagrh, Sundar has been largely sheltered from India's tiger crisis. But had his journey taken him on a different route, his fortune may have been very different.

In the 1970s, Project Tiger was set-up by Prime Minister Indira Gandhi to avert the impending disaster caused by indiscriminate hunting for sport and the skin trade. Then the Indian tiger population had fallen from 40,000 in 1900 to 1,827 in 1972. Three decades later, as Sundar reaches his tenth birthday, India's tigers are once again on the edge: the latest census suggests that India's entire tiger population may number less than 2000 animals. Heavily criticised and lacking the power needed to be effective, in late 2006, Project Tiger was disbanded and replaced by the National Tiger Conservation Authority.

Across India many protected reserves have lost large numbers of tigers to poachers. One 'tiger' reserve, Sariska, was even declared tiger-free in 2005. The true scale of the problem had been hidden thanks to political inactivity and unreliable census figures. A new census by the Wildlife Institute of India, based on estimating prey densities in an area, rather than counting pugmarks has proven much more accurate. The figures announced in 2007 are incredibly alarming and show an overall drop of 60 per cent since the last survey in 2002 (though many individual States dispute the results). The census also indicates that there are very few tigers living outside protected reserves. To make matters worse, many areas have such small numbers of tigers that the populations are considered unviable in the long term.

In 1994, Madhya Pradesh declared itself the 'tiger state', in recognition of its superb natural heritage in Bandhavgarh, Kanha, Pench and other National Parks and of course, its 900 tigers. In 2002, the last 'pugmark' census declared the State had 711 tigers.

Sundar - Three years old

Sundar - Two years old

Sundar - One year old

So it is devastating to discover that the new census estimates that there are now only 277 tigers in Madhya Pradesh.

But, what must come from these new census figures, is not despair or that extinction is inevitable, rather that they should provide the power and determination to make change happen. If decisive action is taken now to combat poaching and to protect the tigers' core habitat, the surrounding buffer areas, and habitat corridors, then there is real hope.

Sundar and the prolific population of the Tala range in Bandhavgarh are a clear example of how India's tiger population could increase significantly in just a few years, given thorough protection. The prospect for the tiger can still be a good one – but action is needed now.

**Organisations working for tigers**

For practical information on what you can do to help protect the tiger and its habitat, contact the following organisations:

**David Shepherd Wildlife Foundation**
61 Smithbrook Kilns, Cranleigh, Surrey GU6 8JJ, UK
www.davidshepherd.org

**Environmental Investigation Agency**
62/63 Upper Street, London N1 0NY, UK
www.eia-international.org

**Global Tiger Patrol**
87 Newland Street, Witham, Essex, CM8 1AD, UK
www.globaltigerpatrol.org

**Wildlife Protection Society of India**
S-25 Panchsheel Park, New Delhi 110017, India
www.wpsi-india.org

Sita F
(born 1982, last seen 6/1998)

Barka M
(first appeared 1984, last seen 1992)

*Born January 1986*
- M (last seen 4/1991)
- F (died 8/1986)
- Hardia F (last seen 6/1998)

*Born September 1989*
- M (killed by poachers 1/1993)
- M (last seen 12/1992)
- F (died 12/1989)

Charger M
(first seen 1990, approximately 5 years old, died 9/2000)

*Born December 1991*
- M (last seen 2/1996)
- M (last seen 4/1996)
- M (died 1992)
- F (died 1992)

*Born February 1994*
- F (died 12/1994)
- Bachchi F (died 3/2

Father - unknown
*Born April 1997*
- **B1** Raj M (last seen 6/2002)
- **B2** Sundar M
- **B3** Barra Larka M (electrocuted 11/2003)

Father - unknown
*Born October 1999*
- **B4** Anuthi F (moved to Khitauli range)
- **B5** F (died in a snare 5/2001)

Father - Sundar
*Born October 2001*
- **B6** M (died 6/2003)
- **B7** M (last seen monsoon 2003)
- **B8** F (last seen monsoon 2003)

Kokila F (last seen 6/2002)

Father - Raj
*Born August 2000*
- **K1** M
- **K2** Juhi F (last seen 12/2003)
- **K3** Chameli F

Father - Shaki
*Born March 2004*
- **C1** M
- **C2** M

Father - Shaki
*Born May 2006*
- **C3** M
- **C4** M
- **C5** F

Father - Barra Larka
*Born June 2000*
- **R1** M (last seen 6/2002)
- **R2** M (last seen 6/2002)
- **R3** M (last seen 1/2003)
- **R4** F (last seen 6/2002)

Father - Barra Larka
*Born April 2002*
- **R5** M (killed by Sundar 1
- **R6** F (last seen 6/2004)
- **R7** F (last seen 6/2004)

Father - Sundar
*Born March 2004*
- **R8** M (living in Michaini a
- **R9** F
- **R10** Tulsi F
- **R11** F

Father - Sundar
*Born January 2006*
- **R12** M
- **R13** F
- **R14** F

# Bandhavgarh's Tiger Family Tree

All the tigers currently living in Bandhavgarh's Tala range, apart from Kokila and Shaki, are descendents of Sita and Charger. Based on their character or where they live, local people have given each adult tiger a traditional name – some have many. The tigresses are often referred to by their territory – Pyari is commonly known as the Chakradhara female.

Since 1997, Kay Hassall Tiwari has been studying the tigers and to avoid confusion, especially when referring to the young unnamed cubs, she devised a coding system based on each tiger's mother's name. The simple code works as follows: the offspring of the tigress Bachchi are B1, B2, B3 and so on. In each litter, males are numbered before females (except Sita's last litter where the females are listed first – Kay started the codes when they were almost adult). Where there is more than one male the numbering follows the sequence in which they were first seen.

Upon reaching adulthood, if the cubs remain in the Tala area, they will receive traditional names and any offspring codes will follow this new name. When Pyari's daughter – P4, established a local territory she became known as Lakshmi. The codes for her cubs are L1, L2 etc.

Key: M - Male
F - Female

*Born March 1996*
- M (died 5/1996)
- M (died 5/1996)
- F (died 5/1996)

*Born September 1996*
- **S3** M (last seen 6/1999)
- **S1** Pyari F
- **S2** Reshma F

Father - Sundar
*Born June 2000*

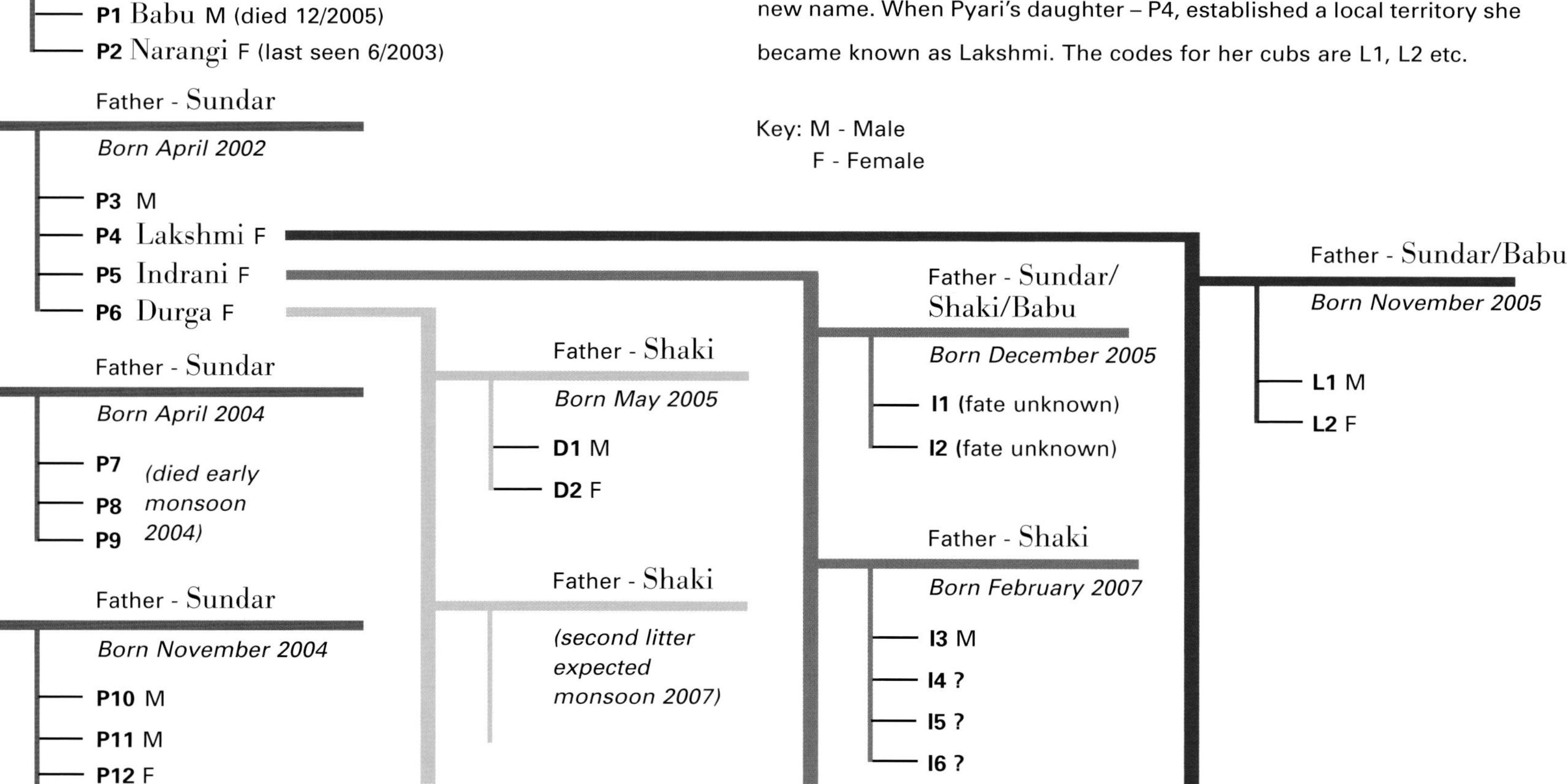

It is the summer of 2007 and against all odds Bandhavgarh's Tala range has more tigers than it has ever had. India's tiger population is in crisis, but Sundar, Reshma and Pyari are ten years-old and are thriving. For the past decade, the jungles around Tala have been their home and despite the huge problems facing tigers across Asia, they have survived and indeed flourished. Collectively, the two sisters have been responsible for more than 25 cubs. Reshma's fourth litter of three cubs were born in January 2006 and during the monsoon they will probably disperse. Two tigers from her third litter, R8 and R10, have taken up residence along the boundary of the Tala and Kalwah ranges. It will be interesting to see whether they stay in the area or move into the adjoining range.

Shortly before the monsoon starts there is wonderful news of Indrani, she is seen briefly with four young cubs, thought to have been born in February. Unusually Pyari's fourth litter are still living within their mothers' range despite being more than two and a half years-old. In the month prior to the monsoon arriving there are no sightings of Pyari; there are two probable explanations. Either she has relocated her territory, so that the females from her fourth litter can establish a range, or more likely she has recently given birth and has tiny cubs. It will not be until after the rains that we shall discover the truth.

The annual rains bring another season of jungle-watching to a close, but Lakshmi and her cubs make one final appearance. In mid-July they were found feeding on a village cow along the main road. Attracting attention for the wrong reasons once again, forest rangers usher the family and their prize back into the forest. During the monsoon, a lack of available territory may force the two sub-adults to disperse – but like India's tiger population, their journey is an uncertain one. Away from the relative protection afforded by the core area of a National Park, can these two tigers find a trouble-free place to live?